Routledge research in environmental politics presents innovative new research intended for high-level specialist readership. These titles are published in hardback only and include:

The Politics of GM Food

'Why have GM Foods become so controversial? Comparing GM food politics in the USA, Britain and the European Union, Toke draws on insights from discourse analysis to help explain this basic political struggle of our time. By stressing the interplay between the material and discursive dimensions involved in the shaping of the conflict, the work offers a detailed account that enriches our political understanding of these "Frankenfoods" on a variety of fronts, in particular the interplay between scientific expertise and citizens' politics. Those interested in the "risk society", both students and specialists, will find much to learn from this perceptive analysis'.

Professor Frank Fischer, Rutgers University, USA

'Genetic modification of food has divided science and politics, regulation and participation, Europe and America, business and the consumer, and values and ethics. In this tumultuous cauldron David Toke has dissected all of the antagonisms to produce a magnificent analysis of a changing world of science, technology, democracy and sustainability that has no equal'.

Professor Tim O'Riordan, University of East Anglia, UK

Why is GM food a controversial issue in Europe but has been given a much easier ride in the USA?

The Politics of GM Food compares and explains how differing political outcomes have occurred regarding GM food and crops in the UK, USA and the EU, thus throwing light on the relationship between science and politics.

Dave Toke uses a discourse approach to analyse the varying regulatory and political approaches, developing a unique framework to describe how different countries have distinctive relationships between scientific assessments of GM food and crops and dominant cultural attitudes.

This innovative volume will interest students and researchers of environmental science, politics and sociology.

Dave Toke is a Research Fellow and Lecturer in Politics and Sociology at the University of Birmingham, UK.

Environmental politics/Routledge research in environmental politics

Edited by Matthew Paterson
Keele University
and
Graham Smith
University of Southampton

Over recent years environmental politics has moved from a peripheral interest to a central concern within the discipline of politics. This series aims to reinforce this trend through the publication of books that investigate the nature of contemporary environmental politics and show the centrality of environmental politics to the study of politics per se. The series understands politics in a broad sense and books will focus on mainstream issues such as the policy process and new social movements as well as emerging areas such as cultural politics and political economy. Books in the series will analyse contemporary political practices with regards to the environment and/or explore possible future directions for the 'greening' of contemporary politics. The series will be of interest not only to academics and students working in the environmental field, but will also demand to be read within the broader discipline.

The series consists of two strands:

Environmental Politics addresses the needs of students and teachers, and the titles will be published in paperback and hardback. Titles include:

Global Warming and Global Politics
Matthew Paterson

Politics and the Environment
James Connelly and Graham Smith

International Relations Theory and Ecological Thought
Towards synthesis
Edited by Eric Laferrière and Peter Stoett

Planning Sustainability
Edited by Michael Kenny and James Meadowcroft

Deliberative Democracy and the Environment
Graham Smith

The Politics of GM Food

A comparative study of the UK, USA and EU

Dave Toke

Routledge
Taylor & Francis Group

LONDON AND NEW YORK

For Ben and James

First published 2004
by Routledge
11 New Fetter Lane, London EC4P 4EE

Simultaneously published in the USA and Canada
by Routledge
29 West 35th Street, New York, NY 10001

Routledge is an imprint of the Taylor & Francis Group

© 2004 Dave Toke

Typeset in Baskerville by Wearset Ltd, Boldon, Tyne and Wear
Printed and bound in Great Britain by MPG Books Ltd, Bodmin

British Library Cataloguing in Publication Data
A catalogue record for this book is available from the British Library

Library of Congress Cataloging in Publication Data
A catalog record for this book has been requested

ISBN 0–415–30699–X

Contents

Acknowledgements

I would like to thank Timothy O'Riordan and Frank Fischer who have given very helpful comments on the book. I would like to thank the many people who gave up their time to be interviewed and the secretaries, Tricia Carr and Lynnete Ford, who transcribed many of these interviews.

I need also to record my great appreciation to the Economic and Social Research Council who paid me the equivalent of a year's salary to research the subject. This was Award Number R00022340. I also want to thank the British Academy for funding research in the USA and the EU.

I would like to thank some individuals who have given me valuable general support. At the top of this list comes Dave Marsh, who co-authored the ESRC grant application for the original research into 'The Politics of GM food' in the UK and who has also been an excellent mentor. Mat Paterson and Dave Humphreys have also given me considerable support and encouragement in recent years.

Introduction

Why have the USA and the EU come to blows over the issue of GM food and crops? How is it that two blocs with such apparently similar levels of technological/economic development and understandings of science produced different outcomes with regard to the application of agricultural biotechnology? In the USA the bulk of canola (oil seed rape) and soya, and much corn (maize) is produced from genetically modified (GM) crops. US consumers routinely purchase food containing genetically modified organisms (GMOs) without much thought or discussion of this fact. Although some GM food crops and food products have been given legal authorisation by the EU, practically no GM food crops are grown in the EU. Moreover practically no supermarket shelf anywhere in the EU can be found containing a product containing more than accidental trace quantities of detectable genetically engineered DNA. These divergent patterns have led to a trade dispute. In May 2003 the USA announced that it was filing a complaint with the World Trade Organisation (WTO) concerning the alleged slowness of the EU in authorising the commercialisation of GM food and crop products.

This book sets out to explain these contrasts. In doing so, however, I want to explore the theory of how science and politics interact (and do so with different effects). I also want to explore the politics of interest group pressure that has been associated with the divergent political outcomes. A 'discourse-analytical' approach is used to achieve these objectives. A social constructionist, discourse approach to this comparative study of GM food politics may be especially relevant to analysis where it is appropriate to assume that none of the contending parties have an automatic right to know and speak the truth. Discourse analysis assumes that truth can only be found in relation to a particular vocabulary and a particular set of values. This approach does not mean that there is no right or wrong, it merely offers an approach which enables us to understand how different notions of truth come to be held with similar vigour by different groups of people.

This discourse-analytical approach is applied in different ways in order to examine regulatory discourses, systems of scientific assessment and advice, and interest group pressure. This analysis is complemented by other techniques. These include examination of cultural and historical context, the use of opinion surveys and also examination of institutional norms.

The epistemic community approach is a well-known way of analysing science-policy interactions. However, I do not use this approach for reasons that are elegantly described in what follows:

> Epistemic community approaches downplay – almost to the point of neglect – the ways in which scientific information simply rationalizes or reinforces existing political conflicts. Questions of framing, interpretation, and contingency are glossed over in an effort to explain politics as a function of consensual knowledge. In failing to consider the nature of discursive practices and strategies, the epistemic communities approach grasps neither the dynamics nor the full significance of intersubjective understandings.
>
> (Litfin 1994: 12)

This criticism is perhaps even more relevant in this case study. This is because I investigate how different sets of actors in different countries have come to sometimes strikingly different 'intersubjective understandings'.

The purpose of this book is analysis of how outcomes have occurred. I do not, consciously, aim to pronounce whether GM crops are or are not in some or all circumstances, good or bad. For me to do so would distract attention from the analysis.

The purpose of Chapter 1 is to explain background arguments that appear prominently in the debate, and to discuss the basic technical and political details of the GM food and crops issue. I begin the chapter by dispelling what I regard as popular myths about biotechnology and the 'world hunger' debate surrounding GM food and crops. Most biotechnology is actually not very concerned with GM food, and despite its appearance in the headlines, the argument about GM food's potential for tackling world hunger seems to be irrelevant to practical regulatory arguments and consumer decisions regarding GM food and crops. I investigate the (changing) attitudes of environmentalists to biotechnology and how it came to be that green groups became more sympathetic to medical uses of biotechnology but not agricultural uses.

Chapter 2 deals with the bulk of the theoretical basis of the book. I discuss the uses and abuses of positivism. I outline the relevance of a discourse approach. I discuss how it is relevant to this case study to borrow from both Foucault's understanding of discourse and Kuhn's notion of paradigms. I call the resulting synthesis a discourse-paradigm approach. This enables me to conceptualise how differing types of positivistic assessment and different attitudes to uncertainty are distinctive to different, nationally based, regulatory regimes. I discuss how the interpretivist, or social constructionist, attitude to knowledge avoids assignation of precise cause and makes the influence of agents and interpretation of events conditional on context. I also discuss the use of network analysis through investigating how different interests ally through support for commonly supported discourses or storylines. I take a look at the theory of the relations between the public and scientific assessment.

I then move on to the case studies. In comparing the USA and the EU, it is useful to take an in-depth view of one country, that is the UK, in order to investigate, in detail, the sort of pressures which have been typical at a nation-state level in the EU. I begin with a discussion of the BSE crisis and its relationship with the GM food issue. Matters such as labelling are set as detailed EU regulations, and so EU member state governments can do little to vary them at a domestic level. However, there is greater discretion over policies on deliberate releases of GMOs in pursuit of the EU's Deliberate Release Directives. Chapter 3 looks in detail at the changing discourses underpinning the operation of the Advisory Committee on Releases into the Environment (ACRE), the main body for regulating GM crops in the UK. The setting up of the 'Farm Scale Evaluation' Trials to investigate the impact of herbicide tolerant GM crops on the food supply for birds is a particular point of attention. I look at how various NGOs associated with what I call the 'wildlife conservation' network have achieved their objectives. I analyse the use of language in forging alliances between the rather disparate set of interests who are critical of GM crops and food.

Chapter 4 examines the case of the USA. I take a look at how attitudes to food may be different in the USA compared to the EU. I also examine how trust in regulators may be higher (compared to the EU) on account of the American public having less suspicion with what industry does with novel technologies. I examine the development of US regulation of GM food and crops and its discursive underpinnings. I examine how the US regulation of GMOs regards genetic engineering as an extension of, rather than a fundamental departure from, traditional plant breeding methods and how some, albeit mainly public relations oriented, changes in regulatory policy have occurred since 1998. GM food and crop policy has been dominated by what is called the 'food chain' policy network. This consists of the main actors in the agricultural and food industry, and this network has enthusiastically endorsed the use of biotechnology in food production. However, cracks have formed in this alliance over developments such as using genetically engineered crops to produce pharmaceutical products.

So why is the EU so different? This is a key focus on Chapter 5 on the EU. I look at European attitudes to food, and the development of legislation on releases of GMOs and labelling of GM food. I look at how legislation and how the dominant discourses underpinning them have changed since the BSE crisis. I discuss the plausibility of explanations of policy outcomes that emphasise this event. I devote a lengthy section to investigation of how the issue has been mediated by the main EU institutions in an effort to assess how typical this issue is in terms of standard theories concerning the working of the EU. I look at how otherwise disparate sets of interests have worked together behind agreed storylines to press their legislative demands.

I have positioned the theory relating to deliberative democracy in Chapter 6 rather than (the theoretical) Chapter 2 for two reasons. First, because it has been suggested to me that it would be better to separate out analytical theory from a normative discussion about how deliberative democracy might best be delivered. Second, I thought it would read better if the theory and practice of public deliberations of GM food and crops policy were placed in the same chapter. I discuss efforts to involve the public in discussions on GM food and crops in the UK, the Netherlands and the USA.

The last section of the book consists of a conclusion. This draws together some answers to key questions concerning the differing political outcomes in the case studies and the development of theory suitable for this analysis.

1 Myths and 'facts' about GM food and crops

Although the central focus of this book is to examine how political outcomes on GM food and crop policy have occurred, this task will be better acquitted if we have a basic understanding of the technicalities and arguments surrounding the issue of genetically modified (GM) food and crops. It is the function of this chapter to achieve this understanding. First I want to examine some myths that surround the GM food/crops debate. Then I shall describe some of what seem to me to be the basic technical issues surrounding GM crops and food. Finally I shall move onto the various arguments surrounding the issue as they appear in the contemporary debate.

The biotechnology myth

A particularly prevalent popular myth concerning biotechnology is that it is mainly concerned with GM crops. It is not. At its most general biotechnology can be defined as the manipulation of biological organisms for human ends. Since even making alcohol using yeast counts as biotechnology under this definition this probably does not really pass muster for a working meaning in today's world, at least going on Wittgenstein's maxim that the meaning of language is in its use. In fact the official UN explanation refers to the notion of 'modern biotechnology' which means the application of:

i *In vitro* nucleic acid techniques, including recombinant deoxyribonucleic acid (DNA) and direct injection of nucleic acid into cells or organelles, or
ii Fusion of cells beyond the taxanomic family, that overcome natural physiological reproductive or recombinant barriers and that are not techniques used in traditional breeding and selection.

(FAO/WHO 2003, Para 8)

As can be seen from the above quote, this definition associates 'modern' biotechnology with genetic engineering. Going on the basis of what biotechnology companies do, this modern biotechnology most often means making medicines and drug testing kits using genetic engineering techniques. The treatment of diabetes has fundamentally changed with the development of modern biotechnology. Insulin was one of the first drugs to be manufactured using genetic engineering, and diagnostic kits derived from genetic engineering are now used to deal with the condition. Larger and larger numbers of pharmaceutical products are manufactured using biotechnology, for instance by using yeasts and other organisms to grow the products. Indeed, at the end of April 2003 'about 370 drugs were in late-stage development or before the FDA (US Food and Drug Administration); for the first time, more than half were developed by biotechnology groups' (Dyer and Griffith 2003). GM food is, by comparison, 'small beer'. Even in the USA, only around 6–10 per cent of all biotechnology companies actually have their main business in agricultural biotechnology (Giddings 2003 – interview).

So, when we see newspaper reports citing employment in European agricultural biotechnology to explain how stunted biotechnology is in Europe, they are making the usual gross over-simplifications that newspapers tend to make. For example, the *Guardian* commented that 'research and development jobs in GM had declined by 60 per cent over three decades in the UK' (Brown 2003). Now, the context of this piece of data was the discussion of GM food and crops, but the received popular impression is that the Europeans are falling behind the Americans in the biotechnological revolution. This is not the case.

In fact there are rather more biotechnology companies in the EU compared to the USA, although the European companies tend to be small, and less well established compared to their American counterparts. According to the European Commission (CEC 2002) European biotech companies employed around 60,000 people in 2001 compared to about 175,000 in the case of the USA. However, the spokesperson for the European biotechnology organisation, EuropaBio, commented that:

> We (the European biotech companies) have got a lot of start ups . . . but I don't think the products are there yet. I guess in essence what happens often is that a small company will be seen. If it develops something useful, (it will) probably get purchased by a larger company. This is the way things go . . . I suspect you may find that a lot of start-ups have actually been incorporated into larger companies in the (United) States once they've got good products.
>
> (Barber 2002 – interview)

The US government has called research into stem cells derived from embryos, while there are fewer restrictions in the UK. It is thus rather debatable whether it is the EU or the USA that is being more restrictive about biotechnological applications. Religiously backed anti-abortion sentiment seems to be rather stronger in the USA compared to environmental concerns, and vice-versa in Europe.

Curiously the company that is most frequently referred to as a biotechnology company, Monsanto, is not a really a biotechnology company at all if one looks at how it earns its money. It earns the bulk of its money from what the chemical industry calls 'plant protection' and what most other people call pesticides. However, because of the demands of being understandable in terms of popular discourse, I do indeed refer to leading GM crop (but in reality, chemical) companies such as Monsanto and Syngenta, as 'biotechnology' companies.

The world hunger myth

In June 2003 George Bush implied that the EU's sloth in approving GM food and crop products was deterring African nations from investing in GM crop technology. According to Bush, the Africans fear that their crops will be banished from European markets. Bush said: 'For the sake of a continent threatened by famine, I urge the European governments to end their opposition to biotechnology' (Goldenberg 2003). Nobel Peace Laureate Norman Borlaug called for the green revolution to be extended to the 'gene revolution' and discussed how biotechnology could help to 'expand the yield potential of crops to improve resistance to insects and disease, resistance to herbicides, nutritional quality and abiotic stresses' (USDA news release 2003a). Crops such as bananas are subject to attack by diseases and there has been research, involving genetic engineering, in countries such as Uganda into ways of combating this problem. Others contend that such problems can be tackled without recourse to genetic engineering.

Certainly, some African countries have been refusing shipments of GM food for famine relief, although in Europe this is interpreted as an American attempt to 'dump' unwanted GM produce. What has struck me is that throughout my examination of the various detailed regulatory arguments about GM food and crops on both sides of the Atlantic I have rarely seen any references as to whether or not GM crops can (or cannot) help solve world hunger. When regulators are shaping rules or considering specific applications concerning GM food or crops they do not, in my experience, ever consider whether GM food or crops will help the poor and starving people in Africa, or even for that matter the poor who are clustered in New York's subway. I doubt very much whether US consumers decide to buy GM food, or European consumers decide not to buy GM food, because GM food will or will not alleviate world hunger. Yet, in general popular debate, this argument features prominently. Perhaps the 'GM crops will alleviate world hunger' storyline serves as a general rhetorical rejoinder to green claims that GM crops are the product of corporate villains who care only for profits.

Whatever the technical possibilities may be for growing high-productivity GM crops in Africa, it is ironic that their development may be hamstrung by the political economy upon which companies like Monsanto rely. In order to develop a GM product a lot of money has to be invested, and private companies are only going to do so if they have patent rights and if there is a good market to supply. The trouble is that starving people, by definition, have no money, and so do not provide a market for the product. Hence agricultural biotechnology companies are unlikely to do much to develop crops for the starving people. Now, the green revolution in the 1960s, which produced hybrid, high productivity crop strains that have been used in Indian and East Asian agriculture in particular, was funded by (largely western) state-sponsored research. Yet state funding to bring products to markets is something that has gone out of favour in the now free-market dominated west. African states have limited resources with which to develop GM crop products suitable for their own food markets.

Greens criticise the green revolution for its dependence on chemicals and its tendency to favour rich owners of large farms who could afford to buy the annual seed requirement. These criticisms are also thrown at GM crops by greens and developmentalist groups. Farmers are dependent on seed suppliers for 'hybrid' 'green revolution' crops because such seeds are, like most species hybrids, infertile. Farmers are dependent on commercial seed suppliers for GM seeds because of patent rights law.

Generally speaking, green and developmentalist groups deny that world hunger is a 'production' problem. They point to the existence of global food surpluses. They say that world hunger is due to distributional problems rather than to a lack of crop productivity.

At the end of the day, GM food and crop technology will be judged by what it can do for the farmers who grow them, consumers who have to eat them, and also on judgements about the environmental acceptability of GM crops. Anything else is mythology. Indeed, even if the USA does win a trade case against the EU, this will do nothing to make European consumers buy GM food. I now want to look at some key background technical factors concerning GM food and crops.

Producing genetically engineered plants

What follows is a brief potted (excuse the pun) account of techniques of genetic engineering of plant varieties and their degree of commercialisation. This is penned, not from the standpoint of trying to satisfy an introduction to microbiology (it certainly would not do that!), but to give some idea of the technical background that is relevant to understanding environmental debates about GM crops. This account is summarised partly from Mayer (2000: 96–99), Stratton (1977), SCP Scientist (2002 – interview) and Merritt (2001 – interview).

The first step is to extract the piece of DNA that has the desired characteristic. The two most widely utilised (in commercial terms) genetically 'transplanted' characteristics have been herbicide tolerance and insect resistance.

Herbicide tolerance means that the GM crop can be sprayed with a broad range herbicide that will kill most types of weeds. The main types of this are ammonium glyphosate, marketed as 'roundup' by Monsanto and ammonium gluphosinate used by Bayer (which incorporated Aventis). Both are regarded by US and European regulators as being reasonably benign compared to other herbicides, although, of course, for fans of organic food, there is no such thing as a benign herbicide. Many types of herbicide tolerant crop are available including corn (maize), soya, canola (oil seed rape) and sugar beet. Herbicide tolerant soya has been the most successful, being grown by the USA and Argentina, the first and third biggest global soya producers. Brazil, the second biggest, is still non-GM as are other important soya producers such as China and India. Herbicide tolerant canola (oil seed rape) is grown in the USA and Canada. Herbicide tolerant wheat has been developed by Monsanto, although, as I shall comment later, farmers are now reluctant to adopt this innovation for fear they will be blocked out of EU and other food markets.

GM insect resistant crops exude a toxin in their pollen that kills insects which would otherwise eat the crop. The insect resistant crops are actually referred to by the name of the bacteria, bacillus thuringiensis (Bt). The most popular Bt crop is cotton, grown in the USA, South Africa, India, China and Australia. Its relatively rapid spread to these countries must be explained partly by the fact that it is not a food crop. This makes the product much less controversial than GM food crops. Bt versions of food crops such as corn (maize) and potato are also available and are widely grown in the USA and Argentina.

Despite the take-up of GM food crops in the USA, Canada and Argentina, the expected spread of GM food technology has not (yet) occurred. People often mistake reports of 'trials' for commercialisation. In China, for example, there have been a large number of trials of a number of GM crops (Huang *et al.* 2002: 675), but, apart from the case of Bt cotton, the trials do not seem to have resulted in widespread commercialisation. Trials do not automatically lead to commercial planting since GM plants which are the subject of trials may not be suitable for commercialisation, they may not be licensed, and they may simply not be grown (as in Europe) because there is no market for their products.

Both herbicide tolerant and insect resistant characteristics are associated with bacteria which can be found in the soil. The required DNA sections are extracted from the chromosomal string of these bacteria by so-called 'restriction enzymes', which are used by micro-organisms to resist attack from viruses. The enzymes act as biochemical 'scissors'. Having separated the vital genes these genes are then spliced onto a 'plasmid'.

Plasmids are self-contained bundles of genetic information which supplement the principal chromosomal bundles held in bacteria, and the desired characteristic is added to the DNA chain in the plasmid by the help of 'ligase enzymes'. These enzymes are normally used in organisms to repair accidental breaks in DNA strands. The assembly of these plasmids are key events, for the plasmids are the agents which carry the required genetic information for implantation into the target plant cells.

At this point it is necessary to mark the bundles of DNA material which are being transferred to the plant in such a way that you can isolate the plant cells that contain the desired new DNA recombination (hence the term 'recombinant DNA'). Herein lies a particularly acute controversy because it was for a long time the usual practice, because of the criteria of sheer convenience, for antibiotic resistant marker genes to be spliced into the plasmids containing the genes being transferred. Genes which confer antibiotic resistance are especially convenient for the purpose of marking the plasmids that contain the DNA that is to be transferred because it is a very simple procedure to isolate the plant cells containing the recombinant by dousing the cell culture with that antibiotic. The cells that do not carry the antibiotic resistant genes die. This appeared to the GM pioneers, to be an elegant way of isolating the cells containing the recombinant DNA. Unfortunately many years later people started to worry that if antibiotic resistant genes were carried into the guts of animals or humans then the DNA might pass into the genes of pathogens. This might promote the proliferation of diseases that were resistant to those antibiotics. I shall say more about this later.

The practice of using antibiotic resistant marker genes in GM plants became a major item of controversy when it involved an antibiotic that was still in use, that is ampicillin, that was used in a variety of Bt corn. Indeed this was associated with a major, and pivotal political crisis in the EU which we shall examine in Chapter 5. In fact other types of marker gene are available today, including genes which allow the cells which have combined with the desired genetic characteristics to grow in the presence of mannose, a type of sugar. In this case the cells that grow are taken for further processing and the unrecombined genes are discarded.

Before this happens the plasmids containing the genes with the new characteristics and the marker genes have to be transferred to the target plant cells. There are different ways of doing this. One way, suitable for 'broad leaved plants (such as sugar beet, soybean and oilseed rape) involves a bacterium *agrobacterium tumifaciens* which is used to transfer the DNA' (Mayer 2000: 97). This bacteria causes crown gall normally, and it is thus quite effective at penetrating plant cells. This is good news for genetic engineers who can place the specially prepared plasmid in a *agrobacterium tumifaciens* bacterium that has been neutered to stop it spreading and it will easily infect the required plant with itself, and with it, the required genes. The cells which have been recombined with the new

DNA will be separated from the rest (as mentioned earlier) and they can then be grown into full sized plants. Then seed can be produced and the product can be tested and, much later, marketed.

Of course a lot of crops are not susceptible to infection by this bacterium, and these include rice and maize. Instead, genetic engineers take recourse to firing gold particles, coated with the plasmids, directly into the cells. This is, as with using crown gall bacteria, a hit and miss affair, but the recombinant genes can be isolated by utilising the properties of the marker genes.

This discussion may serve as a basic description of the GM processes themselves, but some international political processes also deserve inclusion in any sketch of background technical issues relevant to this study.

WTO, Codex and Cartagena

International agreements involving judgements about food health and safety are crucial to the dispute between the USA and the EU concerning GM food and crops. One of the key principles underpinning the World Trade Organisation (WTO) is that rules governing trade must be 'non-discriminatory'. This rules out measures, tariffs or regulations, which favour domestic production over imported goods. States can set strict standards for the environmental performance of goods (for example, motor vehicles), but they cannot, under the WTO rules, set standards for how products were made. If countries feel that other states are breaking WTO rules of non-discrimination, they can make a complaint to the WTO. If the parties involved in the dispute fail to reach agreement then the complainant can convene a Disputes Panel who will make a judgement. Judgements can take a long time to emerge, but when they do injured states are generally given the authority to impose punitive trade measures (such as tariffs) on the guilty parties.

States can protect their consumers' health and safety under WTO rules, although the rules are restrictive about the forms that this can take. The WTO has two agreements which stipulate what is permissible. These are, the Agreement on the Application of Sanitary and Photosanitary Measures, called the 'SPS Agreement' and the Agreement of Technical Barriers to Trade, called the 'TBT Agreement'. These agreements take their standards from a code of practice called the 'Codex Alimentarius'. This code is governed jointly by the UN Food and Agricultural Organisation (FAO) and the World Health Organisation (WHO).

In 1998 and 1999, following EU decisions to label GM food US representatives began to talk frequently about the possibility of complaining to the WTO. The US argument was that the EU's GM food labelling requirements were not in pursuit of genuine safety concerns and were discriminatory against products coming from the USA. Indeed, the USA raised this as a major issue at the Seattle Round of the WTO negotiations in 1999.

However, the USA has received a number of setbacks in its endeavours to stop the EU from labelling GM food.

For a start, an increasing number of countries have decided to label GM food. This includes even those often mentioned by the biotechnology companies as countries 'in favour' of GM crops such as Australia and China. Second, despite resistance from the USA, the Cartagena Protocol on Biosafety, agreed in January 2000, mandates the labelling of imported 'live modified organisms'. The Cartagena Protocol concerns the safety of international trade in live modified organisms. The Protocol is a development of the 1992 Rio Conference's Convention on Biodiversity.

The third, and perhaps the most significant setback has been the drift towards the adoption, in the Codex Alimentarius, of statutes permitting food labelling in connection with foods derived from biotechnology (FAO/WHO 2003, para 19). Codex, of course, is the standard recognised by the WTO itself. Although the USA damns the latest EU regulations, as discussed in Chapter 5, as an extreme and impractical form of GM labelling, it does seem highly unlikely that the USA would win a case against GM food labelling in general. It is significant, therefore, that the USA's WTO case is based on the allegation that the EU has been unreasonably slow in authorising GM food and crop products. It is not based on an attack on labelling GM food.

In fact the US case is only one of several EU/US disputes to be considered at the WTO over recent years. The beef hormone and banana cases were both won by the USA, with the EU giving way on the latter case by scrapping tariffs which favoured banana imports from ex-colonies in the Caribbean. However the EU is itself winning the case over the US imposition of steel import tariffs. In addition, in early 2003 the EU was granted permission by the WTO to impose 4 billion dollars worth of retaliatory trade measures against what were judged to be illegal tax concessions given to major US corporate exporters. If the USA were to win a case on GM food there might be a trade off between the USA and the EU in that both blocs would agree to scrap retaliatory measures. However, if they do not follow this path then there will be major trade war.

Having given a thumbnail sketch of some technical and international political issues, I shall move on to a discussion of the arguments about the acceptability of GM food and crops. What follows below is a summary of some of the arguments in recent public debates about GM crops. As will become apparent from the theoretical discussion in Chapter 2 it is not possible to give an objective account in a controversial topic like this without making clear the values and assumptions which guide this 'objectivity'. The assumption behind this, opening, explanation is that I am covering those 'facts' and arguments which seem, to me, to be most prominent in the public debate in the USA and the EU.

One problem in dealing with the various controversies surrounding GM food and crops is that different interest groups are concerned with different concerns about GM food, or at least they have widely different priorities. A problem for the biotechnology industry is that it is faced with two broad groups of concern; environmental objections and concerns about possible effects on human health. However, radical green attitudes to these issues are coloured by a strategic objection to GM crops. They favour a different trajectory of agricultural development entirely in which GM crops are ruled out almost by definition. Understanding this attitude is very important to an understanding of the debate, although, as will become clearer, the reasons why more mainstream groups, especially consumers, are sceptical of GM food are not necessarily the same as those that drive radical greens. Although the radical greens cannot, on their own, stop GM crops, it does seem unlikely that GM crop technology would have had the problems it has experienced in the EU and other countries without the legitimacy and campaigning organisation that they have provided. So, how did radical greens come, in the first place, to be so opposed to GM crops, and what are their justifications for this stance?

How did greens turn against GM food?

It would be wrong to describe the green movement as a bunch of unswerving dogmatists. They are, in fact, quite pragmatic (or, to their opponents, hypocritical) when it comes to biotechnology. These days greens support medical uses of biotechnology while damning GM crops. Today's greens have changed their tune on medical biotechnology. German greens, as I shall discuss in Chapter 5, campaigned against plans to manufacture insulin using genetic engineering in the 1980s. However, today, greens are usually at pains to give their support to medical uses of biotechnology.

The year 1973 saw the production of the first genetically engineered organism (Manning 2000: 15) and this was followed by heated debates about whether, and under what conditions, genetically engineered organisms (containing 'recombinant' DNA) should be produced. However, the genetic engineering research programme survived some early scares and began to lay claim to significant advances. By the end of the 1970s synthetic human growth hormone and also synthetic insulin had been produced using genetic engineering.

Nevertheless early reactions to the nascent genetic engineering industry from ecologists tended to be hostile and a strong trend of opposition 'in principle' to research into and use of genetic engineering predominated in journals such as *The Ecologist*. In 1983 Jeremy Rifkin, the founder and main force behind the anti-GM campaigning *Foundation on Economic Trends* published *Algeny* (Rifkin 1983). This provided the anti-GM movement with a coherent theoretical framework upon which to base its struggle. I think it makes sense to dwell on the theoretical disposition of Rifkin since he is the world's leading critic of biotechnology from a green perspective.

Although he did not mention Aldous Huxley (or not, at least until the 'revised version' of *Algeny* called the *Biotech Century* (1998) appeared) Rifkin's attack on genetic engineering paralleled Huxley's (Huxley 1950) assault on H.G. Wells's 1920s vision of a technological utopia where man masters nature (Wells 1928). Huxley poured scorn on efforts to 'perfect' the world and humans. Rifkin likened genetic engineering to 'alchemy'. Rifkin explains that just as the alchemist sought to discover how metals can be transmuted to their higher form of gold, a new 'algeny means to change the essence of a living thing by transforming it from one state to another; more specifically, the upgrading of existing organisms and the design of whole new ones with the intent of 'perfecting' their performance' (Rifkin 1983: 17).

In *Algeny* (1983) Rifkin spends a long time attacking Darwin's theory of evolution whose competitive, survival of the fittest, storyline was seen by him to be 'not so much the truths of nature as the operating assumptions of the industrial order, which he [Darwin] then proceeded to project onto nature' (Rifkin 1983: 59) He concludes, after a review of critiques of evolutionary theory that a new cosmology is arising suitable for the cybernetic age which moves away from the Darwinian notion of seeing 'an organism as a concrete structure that performs a specific function' and embraces the idea that each organism is 'a specific store of knowledge gleaned from the larger store of knowledge which pervades the cosmos' (Rifkin 1983: 194–195) This view emphasises how cybernetic and biological information holds the key to the future enabling humans to re-mould their nature and pursue immortality. Rifkin argues that eugenics:

> has quietly slipped in the back door . . . The new eugenics is commercial, not social. In place of the shrill eugenic cries for racial purity, the new commercial eugenics talks pragmatic terms of increased economic efficiency, better performance standards, and improvement in the quality of life.
>
> (Rifkin 1983: 230–231)

Rifkin ends by saying that we face a choice between a bioengineered and an ecological future. 'Genetic engineering poses the most fundamental of question. Is guaranteeing our health worth trading away our humanity?' (Rifkin 1983: 233) He warns that humans will 'devour life around us in order to extend our own' and appears to urge humanity to sacrifice the benefits that bioengineering will bring for the sake of the cosmos (Rifkin 1983: 255).

Rifkin's critique of Darwinian evolution is interesting, although many would see his arguments as being refinements rather than the sort of fundamentalist assault that comes from the surprising number of creationists that seem to exist these days. In fact much of Rifkin's central criticism of genetic engineering reads very oddly viewed from a 2003 rather than a

1983 perspective. He asks whether humans should sacrifice the health benefits that genetic engineering should bring for the sake of the wider cosmos. In contrast, in contemporary debates much of the demands to end genetic engineering in agriculture seem to focus precisely on fears that genetic engineering may harm the health of humans, and also the health of animals and wildlife. Those uses of genetic engineering that are seen as manifestly useful to health have been accepted.

By the late 1980s the focus of debate on genetic engineering had turned increasingly to the topic of GM crops. There had been a number of controversies and court cases (in the USA) concerning the use of genetic engineering in farming, in which Rifkin featured prominently. I shall discuss some of these activities in Chapter 5. For the moment I want to have a look at more contemporary green attitudes to biotechnology. Rifkin's later (or latest) tract on genetic engineering has changed his story, in keeping with contemporary ecological and wider social attitudes. Rifkin argues that:

> It needs to be stressed that it's not a matter of saying yes or no to the use of technology itself ... Rather the question is what kind of biotechnologies will we choose in the coming Biotech Century? Will we use our new insights into the workings of plant and animal genomes to create genetically engineered 'supercrops' and trans-geneic animals, or new techniques for advancing ecological agriculture and more humane husbandry practices? Will we use the information we're collecting on the human genome to alter our genetic makeup or to pursue new sophisticated health prevention practices?
>
> (Rifkin 1998: 233)

This sort of message is now the stock-in-trade of environmental anti-GM food campaigners. For example, Greenpeace's senior EU lobbyist told me:

> It should be clear that our strong opposition is to the use of GMOs in agriculture. We don't oppose biotechnology in medicine. Generally speaking we don't oppose biotechnology when there is no release into the environment by genetically engineered organisms ... I don't want to support biotechnology applied to agricultural, even if there are no GMOs present in the final product.
>
> (Consoli 2002 – interview)

Certainly, there are some clear distinctions apparent between 'medical' and 'agricultural' uses of biotechnology. The emergence of this distinction since the 1980s seems to be heavily influenced by popular opinion, although whether this is seen as praiseworthy pragmatism or just unprincipled hypocrisy depends entirely on one's own agenda!

Biotechnology companies argue that the distinction between genetic engineering and more traditional forms of plant breeding is not clear. The use of genetic techniques to 'mark' genes with desired attributes so that plant breeders can focus their efforts more efficiently is one use that drinks from both traditions. In this case Greenpeace has little difficulty in accepting this as a legitimate, indeed important, use for GM technology in agriculture.

However there is less clarity about the use of radioactive bombardment of genes to produce new strains by chance genetic mutation; that is muta-genesis, which has been used quite often. Mutagenesis has been used for half a century to produce some very important new crop varieties. Popular beer is produced from mutagenetically created grain. A commercially used herbicide tolerant strain of oil seed rape has been produced by muta-genesis and is used in Canada (Dobson 2002 – interview; Consoli 2002 – interview; Merritt 2001 – interview). Such grey areas are used by US regulators to justify their lack of legal distinction between GM crops and crops developed through traditional breeding methods. EU regulators, on the other hand, think that the distinction is usually clear. GM crops are not 'natural'.

By 1987 the campaigns by Rifkin and others to stop open air testing of GM crops seemed to be failing, at least in the USA, and there also seemed to be a groundswell of moderate environmentalist opinion that was favourably disposed towards the new technology. Two questions need to be asked – how is it that environmentalists like Rifkin did not change their stance on GM food as they had apparently changed their view on medical applications of biotechnology? And how is it that environmentalist support for GM food technology is now (so much as it exists) so muted? We cannot determine exact causes, but we can sketch in some relevant background. In answering these questions we can also proceed to sketch out the nature of the radical greens views on how industrial agriculture needs to be 'overthrown', and not reformed with the likes of GM crops.

Elkington and Burke (the latter being a former Director of UK Friends of the Earth) declared, in 1987, that 'a growing number of environmental-ists believe that, in the long term, biotechnology has the potential to solve many of our most pressing environmental problems' (Elkington and Burke 1987: 195) They mentioned Rifkin's efforts to stop deliberate release GM trials but warned that too many hold-ups would result in illegal testing of GM crops and crop treatments. Elkington and Burke pointed to research into a variety of projects including patents for microbes that could clear up oil spills, 'microbes to clean up toxic waste dumps ... enzymes designed to break down wood pulp to ensure that paper mills can extract more useful raw material'. Biological pesticides that are incorporated into plant genes that abolish the need for chemical pesticides and even measures to ensure species conservation were some of the goodies in store (Elkington and Burke 1987: 195–198). So why did these views not become more common among environmentalists? While the biotechnology industry has tended to be very strong, in its promotional materials, on the environmentally sustainable qualities of its GM products and how they will contribute to helping the poor by improving agricultural productivity and also food nutrition, greens have argued that its practice has not lived up to these expectations.

Let us look at a couple of instances here. Way back in 1979 early GM crop entrepreneurs were suggesting that innovations such as improvements in 'nutritional value and (it is hoped) yield, resistance to extremes of climate and soil (such as desert conditions) and tolerance of salinity' would be achieved through GM crop technology. The ability to 'do without expensive phosphate fertilisers' was also suggested (*The Economist* 1979: 72) Unfortunately, very few of the apparently socially desirable objectives have been achieved, so far. In the countries of the South, the only GM crop that has been commercialised on a particulary large scale is GM cotton, which, of course, is not a food crop. At the end of the day if you are left (in the EU) with the biggest potential commercial application being a product that offers a more economical system for applying herbicides, which is what herbicide tolerant crops are essentially about in commercial terms, then you will be judged on this product. This is as opposed to (as I argued earlier) being judged on schemes to help the poor and starving in developing countries that do not yet actually exist.

On the other hand, making GM crops that give consumer (as opposed to producer) advantages, for example, better nutritional value, is rather more complicated than it appears. Not only have you to market the product in such a way to persuade the consumer that the product is good for them but you also have to blend characteristics for the consumer benefit with characteristics that favour competitive levels of yield. Achieving all these things is actually quite difficult. The attempt to market the long-life 'Flavr Savr' tomato in the USA in 1993 foundered, in this case, because the financial operation was misconceived. Some say that people thought the product tasted too 'bland'. These are all very rational reasons why 'consumer quality' GM food products have not been successfully produced, but unfortunately it does not help the biotechnology industry close the gap between environmental and consumer quality performance and its own public relations output.

Of course biotechnologists might have had some hopes that some of the environmental advantages, for example, the avoidance of use of insecticides in Bt crops, might be favourably regarded by environmentalists. However, this has not happened.

There was some initial interest by some organic farmers in technical innovations that might lead to GM crops that could fix their own nitrogen. This could help organic farming which, of course, is based on avoiding reliance on artificial fertilisers. Indeed, as late as 1991, in the UK, a researcher for the (organic) Henry Doubleday Research Association argued (in a Soil Association journal) that: 'The prospect of using at least some biotechnological advances to make organic growing more productive or more profitable may prove difficult to resist'. The comment was also made that 'the biotechnical revolution is being heralded by the same people who brought conventional agriculture to its current state' (Harris 1991: 29).

On the other hand the radical green outlook on agriculture stresses local self-sufficiency and local control of resources. A spokesperson for what became the dominant organic and radical green 'reject GM crops' and 'anti-corporate' position wrote in to counter this view. They argued:

> In general (GM crops) will increase commercial control over the food chain by a few very large companies ... in the case of a herbicide resistance gene, the same company will supply both the seed and the herbicide ... farmers will not retain the right to resow seed or breed from animals carrying a protected gene without permission ... From the perspective of an organic farmer, I find the implications of life patents truly depressing, the exact opposite of the liberating, smaller scale, less intensive, more sustainable future which organic farming holds.
>
> (Goff 1991: 34–35)

Radical greens have come to favour organic farming as the ideal direction for modern agriculture. A key aspect of organic farming (or any system regarded by greens as sustainable) is that farmers control their own resources and reduce the ecological footprints of a given area's consumption. An ecological footprint is the environmental impact associated with consumption, and of course if resources are imported then the environmental impact will be felt somewhere else. Hence the radical green emphasis on areas being, as far as possible, self-sufficient so that the environmental impact is visible and can therefore be controlled so as not to violate the principle of sustainability. The radicals claim that their notion of sustainable, self sufficient, organic farming linked to crop rotation can, with the aid of modern research techniques, help individual farmers maximise productivity and provide a wholesale alternative to conventional farming (Toke 2002a).

It hardly needs to be emphasised that the need to be reliant on a biotechnology company for the supply of seeds runs directly counter to this notion of self-sufficiency. Hence there has been considerable antipathy directed at biotechnology companies by radical greens who see these companies as agents which threaten to throttle their notions of local ecological sustainability. The laws governing definitions of organic farming in both the EU and the USA specifically stipulate that organic food cannot be genetically modified food, and vice versa. This, of course, depends on a political definition of organic food as not being GM. There are no technical reasons why GM and organic agriculture are incompatible, so long as organic agriculture is defined in such a way as not to exclude GM. The distinction between organic and GM food is an example of a social construction that achieves the status of reality as constituted by a dominant discourse. I shall say more about this in Chapter 2. Political factors such as the issue of patenting the exclusive rights to sell certain types of seed to commercial interests, is a vital, perhaps even the most vital factor, that has determined the fate of GM crops in the eyes of many radical environmentalists.

Claims made by biotechnology companies that GM crops are environmentally sustainable are dismissed by the radical greens. The radical greens argue that even if, in some cases, GM crops do reduce the environmental impact of farming on the environment, this is only in comparison to conventional agriculture and not in comparison to the low input or preferably completely organic versions of farming which they promote. Radical greens oppose the strategic objective of trying to reform agriculture through using GM crops in an environmentally sensitive fashion. They see GM crops as merely allowing the industrialisation and intensification of agriculture to continue in a new technological guise using a technical fix that does not remove the underlying problems.

In fact, because GM crops are not infertile they, in a technical sense, do allow farmers to derive seed from the crops themselves. It is the patent laws that prevent this happening. Biotechnologists argue that without patent rights the biotechnology products would simply not be produced. Without a right to receive the profits from GM innovation it would not be economic to match the basic research into genetic manipulation with the right strain and then do all the development and testing. One biotechnologist told me that it cost $4 million to bring a GM crop product to market (SCP Scientist – interview). If the organic sector, for example, does not like GM crops for political reasons, on account of the corporate control over patent rights, then there is not going to be a market for the nitrogen fixing crops that might appeal to organic farmers who want to avoid using artificial fertilisers.

There has been controversy about development of so-called 'Terminator' genes that would make the crops infertile and this is one development that the green movement appears to have deterred. However, because of the exercise of patent rights the technical development is not generally required so long as the patent laws are enforced.

A reformist environmental position is in fact still advocated by the more conservative environmental NGOs, as we shall see in Chapter 3 when I discuss the UK developments and the trials that have been conducted into the likely effect of GM crops on wildlife. Groups like the UK's Royal Society for the Protection of Birds argue that GM crops could be used to benefit the environment. However, the problem here is that the products that have been marketed so far have not been designed specifically to achieve major environmental improvements. The UK government has, at the prompting of wildlife conservation groups, conducted an expensive (and controversial) set of open air trials to examine claims that herbicide resistant crops may damage wildlife.

I want to turn to some of the specific arguments about aspects of GM food and crops which have characterised the 'tactical' battles about GM crops. However, when considering this it is important to bear in mind that the apparent obduracy of radical green opponents to GM crops on specific issues can be explained as being an extension of their strategic objection to the use of genetic engineering in agriculture. First I shall have a look at the effects on use of herbicides and insecticides.

Effects on use of herbicides and insecticides

In the case of herbicide tolerant GM crop regimes herbicides need only (according to the biotechnology theory) to be sprayed once or twice a year. Herbicides like Monsanto's 'roundup' are broad spectrum herbicides which kill practically all types of weed. Monsanto claims that farmers do not, as with conventional crops, need to spray three or four times a year with different types of herbicide to kill the different types of weed, sometimes with herbicides with a much more toxic reputation than those used with GM crops. Atrazine, for example, is used with maize (corn) crops and it has a long residue time in the soil. It has been suggested that it is a carcinogen and also that it is an endocrine disrupter which affects the sex organs of aquatic life and maybe other creatures who drink the water. Biotechnology companies claim that GM crops allow farmers to improve systems of weed control, therefore allowing 'low tillage' agriculture which can, in turn, avoid soil erosion. Low tillage agriculture where fields are ploughed less frequently is normally associated with 'low input' or organic farming. In general the biotechnology companies argue that GM Bt and herbicide tolerant regimes allow farmers to reduce the amount of herbicide and insecticide used to the benefit of the environment and the safety of the farmers.

Anti-GM activists claim that the quantity of herbicide sprayed on herbicide tolerant crops may even increase and that the development of herbicide tolerance among weeds on the farm will make farmers return to using the more toxic herbicides that are supposed to be no longer needed (Cummins 2002). The United States Department of Agriculture (USDA) concedes that the effect of using GM crops on herbicide use depends on the application. According to its research, use of herbicide tolerant soybeans led to small increases in yield, but significant decreases in herbicide use. In the case of herbicide tolerant cotton there were significant increases in yield (and also in financial returns) but no real changes in herbicide use (Fernandez-Conejo and McBride 2000).

In the case of Bt GM crops, the insecticide is carried in the pollen. The bacillus thuringiensis (Bt) is actually used by conventional farmers, including organic farmers, as a 'natural' insecticide. In the case of GM maize (corn), the European corn borer (which otherwise eats the crop) can be killed without use of artificial insecticide. This is also the case with cotton, where the bollworm has become increasingly resistant to insecticides in recent years (Huang *et al.* 2001: 675). Not only do the Bt GM crop systems (allegedly) reduce the time (and money) involved in crop spraying, but because the sprays are more infrequent the farmers are likely to face fewer health risks from crop spraying activities. In fact, as in the case of herbicide tolerant crops, successful application of Bt GM technology appears to vary according to the crop. Critics of GM technology argue that in the case of Bt corn relatively little pesticide is normally used with conventional crops to deal with the corn borer and that consequently little pesticide use is being avoided (Save Biodiversity! 2002). China has apparently abandoned attempts to commercialise Bt corn after 'field tests showed that in China's tightly concentrated farm plots, pests evolved quickly to overcome the resistance of genetically modified plants' (Joseph Kahn, *The New York Times*, October 22, 2002). Even in the USA Bt corn has had a relatively low take-up and the lack of studies published by the USDA on its economic effectiveness is one of those cases where silence can be said to speak very loudly.

Chinese researchers (as well as most other analysts in other cotton-producing countries) report greater success with Bt cotton. However, one extensive study said that while use of Bt cotton reduced production costs considerably, this was because of savings in insecticide and labour costs and not improvements in yield which 'were the same' when 'Bt and non-Bt varieties' were compared (Huang *et al.* 2002: 676).

On the other hand, in the USA, the United States Department of Agriculture has reported that Bt cotton is a considerable success in terms of reducing pesticide use, reducing costs and also through increasing yields (Fernandez-Conejo and McBride 2000). Although the Union of Concerned Scientists has criticised the use of Bt technology on various grounds it also concedes possible advantages:

> Genetically engineered Bt crops offer conventional farmers the option of taking advantage of Bt's safety and efficacy without major changes in their management practice. US farmers are adopting them rapidly where they can realize cost savings compared with current pest control schemes. Most of these savings result from farmers' ability to reduce the number of pesticide applications. In systems that currently depend on multiple applications of synthetic chemical pesticides, e.g. cotton, some farmers can cut their applications of pesticides from six to three. Such foregone pesticide applications represent a considerable environmental benefit for high acerage commodity crops like cotton.
>
> (Mellon 1998: 2)

Effects on wildlife

Ironically, by promoting the notion the herbicide tolerant GM systems kill weeds more effectively, and thus increase agricultural productivity, the biotechnology companies have also caused political trouble for themselves in Europe. This is because there have been fears in the EU that if GM crops kill off weeds more effectively then the supply of food for wildlife, including birds, will be reduced.

The biotechnology companies have a good case when they argue that they reduce the quantity of pesticide used in growing crops as compared to conventional crops, although not, as the radical greens are at pains to argue, compared to organic crops. However wildlife protection organisations in the EU have, since the end of the 1980s, become more concerned about indirect effects of pesticides on wildlife. I shall cover this debate in greater detail in Chapter 3 on the UK. Biotechnology companies have come around, in the EU, to the idea of testing the impact of GM crop systems on wildlife food availability, but the fact that they have been induced to accept this in Europe but not the USA is illustrative of one of perhaps several cultural differences between the EU and the USA concerning agriculture. In the EU most of the countryside is farmed, so what happens to farming very much affects wildlife. In contrast only a minority of the country in the USA is farmed, so wildlife protection is more likely to be seen as something affecting the wilderness.

However, there have been controversies about the effect on wildlife in the USA. The most celebrated argument has centred on research into the effect of Bt GM crops on the monarch butterfly. I shall comment on how this debate has been handled in the US regulatory system in Chapter 4, but here I want to summarise key arguments about impacts of GM crops.

In 1999 a debate on the effects on wildlife or 'non-target species' was sparked after an article in *Nature* concluded that the pollen from Bt corn was toxic to the monarch butterfly, a much loved American insect. In the experiments upon which the article was based, many larvae died when they ate their favourite milkweed leaves upon which Bt pollen had been dusted (Losey *et al.* 1999).

Evidence given to the Environmental Protection Agency (EPA) suggested that only some larvae would be killed and that the concentration of pollen rapidly falls off a few metres away from the corn. The EPA concluded that there were 'no unreasonable adverse effects' on non target species, and no action has been taken to limit the authorisations for commercialisation that have already been issued (Environmental Protection Agency, 2000).

Genetic pollution?

Arguably, the notion of genetic pollution (or cross fertilisation, as the biotechnology language puts it) covers a number of issues. GM herbicide resistant seeds can remain in the soil until a different, successor crop is grown, causing a problem if the farmer tries to kill the weeds to which the 'volunteer' GM tolerant plant is resistant. Biotechnology companies say this can be solved by making plans to ensure that the next crops to be grown will not lead to any problems caused by herbicide tolerance.

Biotechnology companies and conventional farmers who use GM seeds also regard other problems involving gene transfer as 'farm management' rather than 'genetic pollution' problems. One of these problems is simply that the cultivation of herbicide tolerant and pest resistant transgenic crops could, eventually, lead to significant problems from the spread of natural resistance to the pesticides. Biotechnology companies would then respond by producing different versions of Bt to which there was, currently, no resistance. This process has been described as a 'genetic treadmill' (Levidow 2000). Organic farmers, who are generally not well disposed towards the GM crop industry anyway, have become concerned that natural resistance will be developed, by insects, to bacillus thurigiensis. This would make their farming more difficult as the natural spray used to control corn borers would no longer be (so) effective.

A particularly acute controversy concerns the possibilities for cross-pollination and cross-breeding between GM crops and conventional varieties of the crops or, alternatively, wild relatives of the crop species. Possession of the pest resistant pollen seems likely to be a positive asset for plants, and hence it is likely to be passed on where this is possible. The original source of corn (maize) is in Mexico and argument about the extent of the 'threat' to this native gene-pool of corn has become very heated. In an article published in *Nature* Quist and Chapela (2001) claimed that evidence they had collected demonstrated widespread cross breeding between Bt maize and 'natural' maize in Mexico. This, they said, was subverting the genome and posed a threat to the Mexican food supply. Conceivably, if particular crops become dominated by a single genome then they could be susceptible to particular diseases, hence posing a threat to food security. Such dangers are lessened by the use of a variety of strains.

This generated a furious response and, following numerous protests from other scientists, *Nature* took the previously unheard of step of disowning the article. At first the Mexican Government joined the critics of Quist and Chapela's approach, but later tests suggested that an average of 10–15 per cent of Mexican maize crops were contaminated with the transgenic form of maize. The Secretary of the Mexican Government's Commission on Biodiversity described it as 'the world's worst case of contamination by genetically modified material' (Clover 2001). Mayer says that 'Because the majority of today's crop species grown worldwide originated in the South, gene transfers to weedy relatives may be more likely to occur in the developing world rather than in, say, Europe or North America' (Mayer 2000: 106).

Although few seriously doubt that cross contamination like this can occur, there is great disagreement about whether it constitutes a significant threat. Defenders of biotechnology would argue that this type of cross contamination will not make crops more vulnerable to diseases and so the fact that cross breeding is occurring does not matter if there are no major impacts on humans or wildlife.

The spread of genetically engineered traits is also a problem with GM plants modified for herbicide tolerance. In the case of maize grown in Europe the issue will be any gene transfer to conventional maize crops rather than naturally occurring close relatives. However, oil seed rape, a very 'promiscuous' cross-breeder, does have naturally occurring close relatives (brassica) which are normally regarded as being weeds.

Like the 'volunteer' plants of GM seeds which come up in later years when a new crop has been planted, the occurrence of herbicide tolerant weeds is regarded as a 'farm management' problem by the GM industry. On the other hand, ecologists often interpret this as a potential biodiversity problem. A report published by English Nature, the UK's statutory wildlife organisation, which surveyed the North American (especially Canadian) experience, argued that as the number of varieties of a crop species tolerant to different types of herbicide increases, so does the problem of 'gene stacking'. This means that so-called 'superweeds' which are tolerant to two or three (some day, even more) different sorts of herbicide emerge in farmland areas. This could result in farmers resorting to more herbicide use to deal with the problem. This could impact adversely on wildlife if, as a result of 'gene stacking' farmers use extra herbicides on field margins (where birds feed on weeds) in order to deal with the rogue herbicide resistant weeds (Orson 2002). In Canada three versions of herbicide tolerant oil seed rape were in large scale use by 2001. Indeed, the President of the (left-of-centre) National Farmers Union of Canada commented that in the case of GM oil seed rape:

> Far from enabling weeds to be easily destroyed, GM led to the development of more dangerous ones. To remove these new superweeds, our farmers had to turn to old-fashioned, highly toxic chemicals.
>
> (Wells 2003)

However, the biotechnology companies dismiss a lot of the concerns about superweeds as scare-mongering. They regard the problem as a fairly minor problem which can be dealt with by crop rotation. It is argued that if you use a different crop the following year then the weeds that sprout up may be tolerant to last year's herbicide, but not a type used to defend another crop species. Cross breeding occurs only between plants from similar species.

The biotechnology companies also say that the problem will be restricted to farmland where herbicide resistance is an advantage for plants. Moreover, such problems have beset conventional farming for some time without major concern:

> You have oil seed rape that is tolerant to about 20 different herbicides. Now it is tolerant to 23 different herbicides. Maize probably has 30 or 40 products that it is naturally tolerant to. So this whole idea of herbicide tolerance is interesting, but it is adding just one more herbicide that is useful to farmers.
>
> (Dobson 2002 – interview)

There is a more general complaint that the arrival of GM crops, which involves centralised seed distribution, will lead to a loss of traditional varieties. This is certainly an issue in India where eco-campaigners like Vandana Shiva complain about the erosion of biodiversity by modern intensive farming. If particular monocultures become dominant then there may be, as the argument runs, a threat to food security, as mentioned earlier in connection to the row about Mexican maize. However, agricultural biotechnology companies argue that the crop diversity will increase as biotechnology companies keep increased stocks of different crop varieties, albeit managed centrally in their seed banks.

In public terms, the biggest issue concerned with 'genetic pollution' is concern that conventional and organic crops are being 'contaminated', through cross pollination, with GM crops. In this case the concern is not so much proven direct effects but the possibility of as yet unknown effects on human health and, crucially, the wish of many consumers to buy food that is guaranteed to have no GM content. This is especially the case with consumers of organic food who are more sensitive about such issues. Biotechnology companies argue that such fears are unproven and they cite safety clearance by scientific regulatory bodies, including those operating in the EU where consumer resistance has been strongest. Given consumer concerns, labelling GM food products became a demand that the regulators found impossible to avoid, as discussed in Chapter 5. However, mandatory labelling of GM food is still being successfully resisted by biotechnology companies in the USA and Canada. This discussion leads us to a discussion of the effects of GM food on human health.

Effect on humans

Perhaps the most politically explosive issue has been the debate about effects on human health. As we shall see in Chapters 3 and 5 this can be described as the biggest single reason why GM crops have faced so many political problems in the UK and the EU generally. Most would agree that there has not yet been any clear proof of serious ill effects resulting from humans eating GM food. There was controversy over the so-called L-Tryptophan affair, in 1989, when at least 37 people died and 1,500 were left with serious long term disabilities in the USA after an outbreak of 'Eosinophilia Myalgia Syndrome'. The L-Tryptophan, a food supplement, was produced using genetically engineered bacteria. However it is unclear what caused the problem and many have suggested that the problem was the chemicals used in the purification process rather than the GM organisms themselves (Anderson 1999: 17–19).

There have been some instances where trials have revealed potential for adverse health effects, but such instances are variously interpreted as evidence of problems with GM food or evidence of the effectiveness of the safety assessment system for GM food. The most well known instance of experimentally generated concern was the set of experiments conducted by Dr Pusztai in the UK. His research was publicised on a TV programme in August 1998. Pusztai fed a type of genetically modified potatoes to rats and reported that the rats became ill as a result. This affair was interpreted according to people's prior belief systems. To some (including the UK's prestigious Royal Society which represents leading scientists) this was a faulty experiment, to others it was cause for concern. Pusztai became a cause celebre, especially because his employment at the Research Institute was abruptly terminated following the TV programme.

Since then Pusztai has expounded the reasons for seeing genetically engineered food products as being clearly different. He argues that there are uncertainties about their effects on human health (Pusztai 2002). In general the ingestion of novel DNA fragments should not, logically, pose any special risks given that, for example, just because one eats cabbage does not mean that one will sprout leaves. However, Pusztai has suggested that the cauliflower mosaic virus that is routinely used to 'switch on' the gene that has been transferred is subtly different from normally consumed (eaten) viruses of that type. He says that this could cause unintended alterations in the genetic makeup of the modified plant. He claims that '[T]he top scientific establishment ignore these uncertainties, perhaps because they are reluctant to fund research to investigate them' (Pusztai 2002: 83) Various radical critics of GM food technology have pointed to ways in which GM food differs from the conventional varieties because other genetic material, besides material which relates to the specific attribute that is being conferred, is also transferred with unknown consequences.

The fact that there is no manifestly clear link between GM food and ill effects in humans has not deterred GM critics from emphasising the risks. Faced with the claim that US citizens have been eating GM food since 1996 without apparent health problems, the Greenpeace's EU lobbyist on GM foods said:

> You know how many years the British have eaten mad cow meat before they started to talk about new variant CJD? Mad cow disease started before 1984 (but human effects were not noticed until 1996).

When you have long term effects you cannot predict them ... In Europe after comprehensive labeling after 10 years you may be able to say something about this after serious epidemiological studies, which you cannot do without control groups.

(Consoli 2001 – interview)

Critics of agricultural biotechnology argue that it is possible that new allergenic reactions can be produced by eating food with novel combinations of genes. A report produced by the UK Royal Society commented that 'The allergenic risks posed by GM plants are in principle no greater than those produced by conventional or organic means' (Royal Society 2002: 8).

Nevertheless, genetic transfer with bacteria is commonplace and particular concern has occurred over the effects of eating GM food containing antibiotic resistant marker genes. As mentioned previously, earlier versions of GM crops (although some are still in use) used antiobiotic resistant marker genes which led to fears that this could spread, in stomachs, to pathogens which could themselves become resistant to antibiotics. Despite the fact that scientists said this was unlikely, this seemed too great a risk for many people. Increasingly, regulatory institutions have demanded that GM crops use different types of more benign marker genes. As we shall see in Chapter 5 the use of ampicillin in the first type of transgenic maize that was imported into the EU proved to be a spectacular political own goal in the EU.

Future developments

Judging from current trials going on in the USA, future developments are likely to occur in four directions. First, development of more crop strains that have advantages for farmers such as crop varieties which are resistant to various types of diseases or which are suitable for currently inhospitable conditions like high salinity, low rainfall or excessive soil aluminium. Biotechnologists point to the adoption of virus resistant GM technology which they say has saved the papaya industry in Hawaii. Second, crops which deliver 'quality' attributes to the consumer like the high beta-carotine, Vitamin A inducing, (and much talked up) 'golden' rice being developed in India. Third there are 'plant based pharmaceutical crops' (or biopharming) that can be used to grow medical products like blood thinners, clotting agents, anti-arthritis, contraceptive products or even anti-cancer drugs. Fourth are animals modified either to deliver larger amounts of product, like the large salmon that are already being considered for commercial approval in the USA or animals that have been genetically modified to produce drugs for humans. For example, the widely used enzyme-drug trypsin can be produced from genetically engineered animals.

Some of these developments, notably plant based pharmaceuticals and GM animals, are already quite controversial in the USA where the development is most advanced, which does not bode well for their acceptance elsewhere. I shall touch on some of these controversies in Chapter 4.

Some nature conservationists question the automatic assumption that genetic engineering designed to help crops grow in unhospitable circumstances will be ecologically viable. For example, the biotechnology spokesperson for English Nature, the UK's wildlife watchdog, criticised plans to produce GM crops that can prosper in salty soil. He commented:

> Has anyone questioned the wisdom of continuing to grow crops in regions where soils have been ruined by centuries of overgrazing and inappropriate irrigation? ... By developing plant varieties that allow damaging agricultural practices to persist we run the risk of exacerbating existing soil and salination problems, and may increase the demand for irrigation in areas where water resources are growing scarce ... The search for sustainable food supplies in arid saline areas is primarily about finding sustainable agricultural strategies, and should not be sidetracked by debates about plant-breeding methods.
>
> (Johnson 2003)

Certainly the downbeat assessments made by many 'establishment' as well as radical environmentalists in Europe provide a stark contrast to the rallying cries for the cause of biotechnology made by the scientific elite in the USA. As a keynote paper at a National Academy of Sciences colloquim put it: 'Widespread adaptation of biotech-derived products of agriculture should lay the foundation for transformation of our society from a production-driven system to a quality and utility-enhanced system' (Kishore and Shewmaker 1999: 5968).

It should be borne in mind that we are comparing divergent scientific opinions here. As we shall see in later chapters the contrasting EU and US approaches permeate their respective regulatory systems and modes of scientific assessment of GM crop and food technology. In Chapter 2 I shall develop the 'discourse-paradigm' approach which supplies a theory which enables us to analyse the distinctive patterns of scientific and cultural approaches to GM food and crops that are apparent in different countries and regimes.

Theoretically the biotechnologists will be on relatively safe ground with products where there is no fear about the consequences of 'genetic pollution'. For example, will there be any tears if the trait giving higher levels of beta-carotene spreads to other types of rice? Perhaps not, but many anti-GM groups are still opposed to this innovation arguing that as yet unknown genetic modifications will enter ecosystems.

However, as far as future development is concerned, the biggest problem faced by agricultural biotechnology is consumer resistance. This phenomenon is studied and its political consequences are analysed in the chapters on the UK and EU. Unfortunately for the biotechnology industry, European consumers may prove not to be idiosyncratic in their scepticism about GM food. In India, where the dominant Hindu religion favours vegetarianism and where many are worried about animal genes being spliced onto vegetables, sale of GM food is still, at the time of writing, actually illegal. Biotechnologists have long held hopes that China would prove more receptive to GM food technology, but commercialisation of GM food technology seems to have stalled (Joseph Kahn, *The New York Times*, 22 October 2002). Increasingly farmers around the world are becoming unwilling to grow GM crops lest their produce becomes unsaleable in a rising tide of consumer resistance to GM food. In short, the USA may win some sort of victory at the WTO, but the key battle is in the food markets, and here it is heading for defeat.

Conclusion

We can see from the description in this chapter how suspicion of genetic modification has a longstanding cultural pedigree in the Anglo-Saxon world. Notions that it was man's destiny to perfect nature and even the human form according to scientifically determined laws of efficiency were first contested and then, apparently, rejected in favour of a more cautious approach. Warnings made by alleged or even self-confessed romantics have proved to be highly influential in comparison to self appointed high priests of the new religion of science. This cautious approach neither rejects nor automatically accepts technological innovation which alters our relationship with nature. It accepts that there is an inherent risk-benefit evaluation to be made.

In the 1970s and 1980s radical greens, led by Rifkin, seemed to oppose any adoption of genetic modification techniques in order to avoid the subversion of both humans and ecosystems by a technological devil. However the greens have swam with the tide and accepted medical applications of genetic engineering where the benefits are perceived by a social consensus to outweigh the risks. This approach seems to break down when it comes to agricultural biotechnology. Here radical greens stand fast against virtually anything that involves the release of genetically modified organisms into the environment. Arguably, they have been allowed to continue on this course simply because the GM crop industry has not seen fit to guide its activities into areas which a European, as opposed to American, social consensus deems to be overwhelmingly good for the consumer and/or the environment. We shall investigate the differences in US and European approaches in later chapters.

The arguments posed by biotechnology companies that GM crops will help solve starvation problems in the Third World have considerable appeal. However, it is not clear how the growing of herbicide tolerant crops in European countries will achieve this objective. Moreover the biotechnology companies' control of patent rights for seeds seems to distort the direction of biotechnological research so that they investigate what will make money in rich countries rather than what will help poor countries where the problems are lack of money rather than (merely) a lack of food. If the conquest of world hunger is a major selling point for GM food crops then one would have expected the technology to have made much more rapid progress in developing countries (such as India and China) than it has. Of course GM (Bt) cotton has (so far) proliferated in several countries, but this is not a food crop.

One feature of the GM crops and food issue is the difficulty of assessing claims and counter-claims, as is apparent in the previous description. Only part of this can be put down to over-enthusiastic blurring of facts and propaganda on such as issues as whether GM crops do increase yields or whether there really are instances where people have been harmed through eating GM food. However, a lot of the problem is that it is difficult to draw a line between science and politics, however hard people try to claim that they can do this.

Perhaps the most fundamental aspect of this story is how science and politics have interacted. However we cannot say much that is meaningful about this unless we discuss how this occurs. We have, for example, spent a long time in this chapter talking about how the role of science is perceived. Perhaps it is the way that science is perceived, and the value systems that underpin systems for scientific evaluation that are the key aspects to understanding this story. In the next chapter I shall develop an approach to analyse how scientific regulatory systems have dealt with the GM food and crop issue.

2 From science to politics

The principal function of this book is to study how science-oriented regulatory systems in the USA, the UK and the EU have come to choose rather different bases upon which to assess and order GM food and crops. That is an analytical question. This chapter discusses some theoretical tools to enable us to engage in the analysis. The argument about the role of science in the GM food and crops debate does have a lot to do with previous discussions of the science and politics of environmental and health regulation. Jasanoff (1990) talked about how the debate about the role of science in regulation had been polarised between those arguing from a 'democratic' viewpoint who saw scientists as making a distorted contribution to regulatory outcomes and those arguing from a 'technocratic viewpoint' who argued that only scientists made, or should make, regulatory judgements. The democrats tended to argue for greater public debate and participation, while the technocrats argued that decisions should be left up to the relevant scientists.

Jasanoff concluded her study of the role of scientific advisers in the US regulatory system by arguing that both the 'democratic' and 'technocratic' stories were simplistic. It was wrong to suggest that scientific judgements could be divorced from social, political, industrial and cultural influences. However, it was equally wrong to suggest that scientists were somehow irrelevant to the decision making system. This is because the arguments were conducted in, or translated into, a scientific language. Scientists play vital roles as interpreters, even if what are essentially arguments about values are conducted in scientific language.

Others such as Liberatore suggests that 'Scientists are socially situated reasoners rather than bearers of universal truth' (Liberatore 1995: 61). I would concur with this statement, but I would also argue (lest there be any misunderstanding) that scientists are necessary contributors to deciding the social truths of the day, even though we ought to recognise that such truths are contingent, that is specific, to some societies at particular times.

The US attacks on the EU's stance have a technocratic ring about them in that the US Government appeals to the decisions of scientists. Organisations like the US Food and Drug Administration proclaim that their approach is 'science based'. Three issues emerge here. First we shall see in succeeding chapters that in the case of some issues, officially appointed scientists in EU member states have had distinctly different approaches to GM food and crop technology compared to officially appointed US scientists. Second, the claims that any regulatory system is based solely on positivist (that is, experimentally-based) judgements, as opposed to merely involving the advice of some scientists, is highly dubious. Third, the notion that science itself can be divorced from cultural values and set up as some sort of diviner of universal truths about social or regulatory issues is also highly dubious. I shall leave it to later chapters to deal with evidence that deals with the first issue. In this chapter I take issue with the second and third notions from a theoretical basis. I set out a mechanism for analysing and comparing regulatory politics in different regimes based on a mixture of analysis drawn from Foucault and Kuhn and some other analysts who have used their ideas. I then move on to discuss an approach which is suitable for explaining how political outcomes have emerged in the governance process and the role of the public in what are seen as scientific deliberations.

I begin this theoretical work with a critique of positivism. A critique of positivism can help us understand how some may challenge the legitimacy of USA, or for that matter, European, claims to base their regulatory decisions on 'scientific evidence'.

Some uses and abuses of positivism

In general, positivism is a means of legitimising statements about the nature of the world that may be useful to us. Positivism can mean something narrow or broad. Narrowly it refers to classic scientific methods of subjecting hypotheses to experimental tests in order to judge their validity. This is done by careful design of the experiment(s) in order to control all the variables except the one(s) under inspection. Statistical analysis is conducted to estimate the probability that the results have occurred by chance. This method is generally regarded as the most rigorous means of collecting scientific evidence. The general methodology itself is not controversial in contemporary discourses on both natural and social sciences. What is more controversial are the claims that positivists make for their methods. Positivists claim that their methods reveal truths about the nature of absolute, universally applicable, reality. They may also make claims (with more obvious political overtones) that establishment scientific regulatory regimes of one sort or another have a monopoly on truth statements about the things they regulate because they accept the word of the scientists who conduct the relevant experiments (Fischer 2003).

One leading British biotechnology research scientist argues that 'We must ensure that safety evaluations are based on science and scientific data alone. If political and other factors must be brought into consideration, those responsibilities must lie with the appropriate expertise else-where' (McHughen 2000: 631). This is a broadly positivist statement which assumes that we can, and do, make regulatory judgements on the basis of scientific data alone. The claims made by the US government on GM food are also positivist. They say that their regulatory system is 'science based', that it has given the all-clear to GM food and crop products through scientific means.

A couple of problems emerge here. First is the fact that most GM foods sold in the USA today have not been the subject of scientific assessment in the sense that there have been exhaustive tests about whether they are safe. As we shall see in Chapter 4 it has been assumed that they are safe because companies have notified the Food and Drug Administration that they are similar to conventional foods which are 'Generally Regarded As Safe' (GRAS). What is meant by safe? It depends on prevailing cultural values which may, one day, decree (especially in a more vegetarian world) that 'ordinary' bacon is unsafe because it contains animal fats! It depends on what the public generally refer to as 'common sense', that is common views of what is true, not scientific research and experiments in themselves.

Hence the notion of GRAS is a social construct. Even if one could test everything and rule what foods are safe, or not, according to an agreed standard, there would still be no end of argument about what constitutes a healthy diet. Presumably tomatoes are 'safe'. But is it safe to exist solely on a diet of tomatoes? Most people would say no, or would at least prefer not to! Of course, even though the notion of GRAS may be a social construct, it is still reasonable to point out that foods such as tomatoes which have been the targets of genetic engineering, are indeed generally regarded as safe, provided of course you eat other things as well!

However, this still leaves open the issue of whether GM tomatoes are 'the same' as non-GM tomatoes. That, however, is still a value judgement that is difficult to validate through any positivistic experiment. Sure, one can demonstrate (as is done in the USA) that various food properties of GM and non-GM soya are the same. However, people can still point out that the DNA is different, and it becomes a matter of judgement (prejudice?) as to whether possession of different DNA conferring some attribute (for example, herbicide tolerance) can actually ever make any notable difference to food quality.

Now I am not trying to use fancy arguments to question the safety of GM food per se. What I am trying to do is to say that however strongly many people believe GM food to be safe, they are not reaching this conclusion because of scientific experiments; they are doing on the basis of belief based on confidence in what regulators say, what official scientific advisors say, on what industry does, what people say in the newspapers, what their friends say, their 'gut feeling' and so on. Experiments can help us form these judgements, when they are carried out. Ironically, one of the complaints made by many GM critics is that the experiments have very often not been carried out enough.

A second question is the issue of what scientific experiments to conduct. The Americans, it seems, construct the possibility of GM crop impact on wildlife as being represented by the effect on butterflies. On the other hand the British have constructed wildlife impact as the effect on food supplies for birds. So which experiment produces truth about environmental impact on wildlife? The positivist may say that this is a political, not a scientific, judgement. Well, that is the point. What positivism can actually deliver is much narrower than the claims to be a portal on absolute reality that one hears. 'Let science decide!' is the call. Yet the issue of what experiments should be conducted, the issue of how the results should be interpreted and, of particularly strong importance, how uncertainty should be interpreted are all matters that cannot be adequately settled by the positivist methodology.

Levidow, echoing a point made by Wynne, says that 'scientific uncertainty cannot be properly described as objective shortfalls of knowledge. Rather the perceived uncertainty is a subjective function of complex and cultural factors' (Levidow 2001: 845; Wynne 1992a: 120). The precautionary principle, a contested notion concerning relationships between risk and uncertainty (O'Riordan and Cameron 1994), becomes a focus of debate in cases where a case study is thought to involve scientific uncertainty.

Habermas, in one of his early works *Towards A Rational Society* (Habermas 1971), argued that positivism was used by military-industrial groups as a cloak to limit public debate and reserve decisions to an agenda which they dominated under the guise of technocracy. More recently, analysts such as Wynne (1992b) argue that 'there is no preordained and objective definition of the boundary between science and politics'.

I agree with this. Indeed this implies that science itself cannot be divorced from cultural, social and political influences. Even when we are talking about experiments to demonstrate matters of pure natural science (never mind the much more politicised world of regulatory science) there is a lot of evidence to suggest that it is not experiments, themselves, which tend to confirm or disprove theories in the minds of the relevant scientific peer-group. Using a series of key historical examples (Collins and Pinch 2000) demonstrates how belief systems influenced the acceptance of new theories about the world rather than the results of experiments whose results were wide open to different interpretations. Kuhn (1970) came to much the same conclusion about the factors which led to the acceptance of one new type of theory (paradigm) over another. Scientific tests are selected and the results are interpreted according to values which are culturally derived. Some well known theories are not even capable of generating testable (and as Popper explained, falsifiable) hypotheses. Strictly speaking, the fact that the theory of evolution does not generate testable hypotheses consigns this theory to the status of the unfalsifiable and therefore philosophically dubious (Huxham 2000: 13). Of course Darwin conducted a lot of careful research, but belief in the relevance of his theory rests on cultural grounds. It resonates with 'modern values' that do not take biblical texts literally. As we saw in Chapter 1, there are disputes about how the theory should be interpreted. Ordinary culture, that is values and social truths held dear by the population in question, is part of science, and science is part of culture. Bruno Latour explains this.

Of course, it may be, as the USA alleged, that the scientific evaluation of GM food and crops in the EU has become bogged down in simple pig headedness despite scientific evaluations. However, regardless of this issue, there are, as the following chapters suggest, some clear differences in scientific evaluations and regulatory procedures in the EU and UK compared to the USA. We cannot analyse the issue simply in terms of how one side in the transatlantic dispute has or has not got it right. We have to analyse how the different outcomes have occurred.

In order to analyse how different outcomes have occurred I want to adopt several techniques, but a central part of my method of understanding the differences in scientific evaluation between the USA and the UK/EU is an adaptation of the discourse and paradigm approaches to scientific knowledge.

Discourse and paradigms

Foucault and Kuhn both studied sciences, albeit of different types, and came to conclusions that were, and still are, unsettling to many, especially to positivists. Foucault and Kuhn both argued that scientists were not purveyors of absolute truths who had privileged access to reality. Rather that scientists used systems of reality, ontologies, that tended to change as time, technologies, social requirements (especially for Foucault), required. These systems of thought, 'paradigms' for Kuhn and 'discursive formations' for Foucault, represented incommensurable worlds where categories and standards could not be translated into each other. These discursive formations or paradigms did not bring science closer to the truth as they succeeded each other, they were just different. Kuhn said 'We may . . . have to relinquish the notion, explicit or implicit, that changes of paradigm carry scientists and those who learn from them closer to the truth' (Kuhn 1970: 170).

There were also some clear differences in emphasis. For a start, as Gutting (1989: 3–7) points out, Foucault looked at what he saw as 'dubious' human sciences like psychiatry, medicine, law and sexuality. Foucault's rejection of these science's truth claims was more scornful than Kuhn's. He was especially dubious about these sciences because he saw them as means of social control, not as a means of dispassionate study of the nature of human beings. Foucault used his analysis to debunk the omniscience of established orders, which is reasonable enough, but then implied radical political conclusions. Indeed this apparent radical, some say anarchistic, function does, I feel, distract attention from some of the highly useful analytical features of his approach. Foucault focused on discourse and language, and later on the relationship of discourse to power and knowledge. He implied that these forces acted to constitute the individual rather than individuals having power to freely organise their lives.

Kuhn, on the other hand, concentrated on the process of changing scientific paradigms and the conditions under which what he called 'normal' science was conducted. He studied natural sciences, especially physics and chemistry. He said that paradigms consisted of the entirety of belief systems associated with a particular theory and also the exemplary experiments which were held to demonstrate aspects of that theory. Kuhn studied the way that scientific theories developed, how scientific revolutions occurred and how, in between scientific revolutions, 'normal' paradigmatic science was simply a process of 'puzzle-solving', of filling in gaps in the knowledge structure of the paradigm. He was willing to support a notion of scientific progress:

> Later scientific theories are better than earlier ones for solving puzzles in the often quite different environments to which they are applied. That is not a relativist's position, and it displays the sense in which I am a convinced believer in scientific progress.
>
> (Kuhn 1970: 206)

I find it useful to take some techniques from Foucault (and some analysts of environmental policy who have worked in his name) and also some techniques and concepts from Kuhn. I shall begin with Foucault's approach. Foucault is associated with discourse analysis, although that in itself can mean many things. My purpose here is to describe my use of this technique rather than to explain the range of other uses of the technique. I take a cue from Howarth (2000: 131) who says that 'discourse theorists are concerned with how, under what conditions, and for what reasons, discourses are constructed, contested and changed'.

A major advantage of the Foucauldian technique of examining discourses in order to reveal practice is its potential for analytical neutrality in the sense that the focus is not distracted by speculative assessments of the motivations, effects and interests of particular agents. Power, for Foucault, is not a property of particular agents, it is a relationship that thoroughly permeates all segments of society and is intimately bound up with the collection and legitimisation of certain forms of knowledge.

For me, a key value of discourse analysis is that it allows the analyst to treat different conceptions of reality (for the purposes of analysis) as being equal in terms of their truth value, regardless of the claims of particular agents. Hajer states that:

Discourse is here defined as a specific ensemble of ideas, concepts, and categories that are produced, reproduced, and transformed in a particular set of practices and through which meaning is given to physical and social realities. The argumentative approach conceives politics as a struggle for discursive hegemony in which actors try to secure support for their definition of reality.

(Hajer 1995: 44 and 59)

Litfin adopts a similar strategy and, summarising Foucault's approach, argues that 'What becomes important, then, is how certain discourses come to dominate the field and how other, more marginal counter-discourses establish networks of resistance within particular "power/knowledges" or "regimes of truth"' (Litfin 1994: 38 citing Foucault 1980). Foucault himself pointed out that 'it is in discourse that power and knowledge are joined together' (Foucault 1998: 100).

These quotes imply that, in order for us to chart how certain discourses become dominant (achieve hegemony), we have to do so from a basis of identifying and analysing key discourses which underpin the regulatory arrangements that we are studying. By way of one relevant example, a discourse legitimising power and knowledge occurs when a dominant regulatory discourse states that a particular environmental impact is important, which thus implies a particular power relationships between say, biotechnology companies and some environmental groups. A particular discourse will sanction certain experimental tests, or the opinion of certain precise authorities, as legitimate relevant knowledge. By the same token, by implication some groups will be sidelined and some sources of knowledge will be marginalised or rendered irrelevant for policy and political purposes. In this sense discourses can act in the same way as Kuhn's paradigms do in legitimising certain types of scientific knowledge and experimentation and excluding others.

Discourses, however, are not like paradigms, which are coherent and essentially immutable in form, in that discourses can be made up by changing the arrangement of an almost infinite number of combinations of elements. 'We must conceive discourse as a series of discontinuous segments whose tactical function is neither uniform or stable' (Foucault 1976: 100). A clever Machiavellian Prince (or alternatively a British Minister for Environmental Protection) will re-write the dominant discourse. They can do this by adopting some elements of rival discourses in order to bring on side some forces whom he/she can safely accommodate, while leaving out those elements (and associated forces) that he/she finds least conducive for his/her purposes. We thus need to study how dominant regulatory discourses change, or even, in some cases, how they do not change despite apparent pressures. There are plenty of examples of these tendencies in the succeeding chapters.

My usual focus for sources for dominant discourse in this work are regulatory texts, rules, justifications for rules, reports of regulatory committees and such like. I also use statements by agents who can be regarded as being sources of knowledge which have been legitimised by dominant regulatory discourses. There is always going to be an argument about what quotes you include, but I think it important to focus on key passages which have been the focus of the keenest debate and which most pithily encapsulate the identity of what distinguishes the dominant discourse from other discourses. These dominant discourses are more than mere policy statements and justifications for these policy statements. They are, as explained earlier, significant as sources of the power and legitimised knowledge inherent in the policy area with which they deal. However, my selection of discourses for analysis includes a major focus on scientific as well as legal interpretations.

I argue that the GM food and crop regulatory discourses which I am discussing share some of the 'discontinuous' aspects of a discourse and, when added together, some of the natural scientific aspects of a Kuhn-like paradigm. Hence I use the term 'discourse-paradigm', to describe the sum total of legal and scientific regulatory discourses about GM crops and food in a particular state. These discourse-paradigms can be used as the basis for comparisons between one country and another.

These discourse-paradigms are related to distinctive national sociological and political idiosyncracies which persist over time. It is true that Foucault himself developed the notion of 'discursive formation' precisely to deal with the issue of how human-scientific disciplines were organised over a long period of time (Foucault 1994). However the idea of discursive formation seems to be concerned with developing an underlying theoretical linguistic structure more than something that can be applied to discussions about the practical fact collecting activities of regulatory science. This is a reason for the inclusion of the term 'paradigm' in 'discourse-paradigm'. However there are additional reasons for using the ascription 'paradigm'.

The type of activity practiced by natural scientists who work to generate or justify knowledge associated with a particular GM food and crop regulatory regimes is analogous to the 'puzzle solving' which Kuhn saw as the main activity of 'normal science'. This 'normal science' is said, by Kuhn, to be the type practiced, in usual times, when a particular paradigm is dominant. Kuhn explains:

> [N]ormal science ... seems an attempt to force nature into the preformed and relatively inflexible box that the paradigm supplies. No part of the aim of normal science is to call forth new sorts of phenomena; indeed those that will not fit the box are often not seen at all ... Instead, normal scientific research is directed to the articulation of those phenomena and theories that the paradigm already supplies.
>
> (Kuhn 1970: 24)

Kuhn described the activity of normal science as being concerned with the collection of various types of fact that are deemed important by the paradigm (Kuhn 1970: 25–27). This type of fact-collecting is analogous to the fact collection which is demanded by GM regulatory discourses on GM food and crops, whether it is information on whether GM food is 'substantially equivalent' to conventional food or information about different types of environmental impact caused by GM crops. Different discourse-paradigms will demand different types of information and the organisation of different sorts of experiments.

Kuhn attaches considerable importance to the scientific community as the basis for legitimising paradigms. In particular the 'group that shares ... (the solutions to natural scientific problems under discussion) ... may not ... be drawn at random from society as a whole, but is rather the well-defined community of the scientist's professional compeers'. Kuhn also rules out appeals to 'heads of state ... or to the populace at large in matters scientific' (Kuhn 1970: 168) as the basis for legitimising the paradigm.

The paradigm concept has an important lesson for this work in focusing attention on the composition of the scientific bodies which make up the scientific regulatory committees. However, Kuhn's discussion of scientific communities also appears to signal an important way in which his 'paradigm' approach may, at first glance, not be compatible with the study of GM crop and food regulatory discourses. This is because the composition of the scientific regulatory committees is designated precisely by the political mechanisms which Kuhn proscribes. However, before agreeing this caveat in my attempt to borrow from the paradigm concept I would want to consider the attitude of the wider scientific community to the key aspects of specific examples of GM food and crop regulatory discourses.

Does the wider scientific community in, say, the UK and the USA generally support the various discourses that constitute their GM food and crop regulatory system? If they do then the Kuhnian notion of a scientific paradigm is preserved in the sense that the paradigm is selected by the scientists, and not (just) the government. On the other hand if the balance of scientific opinion (as expressed through the official scientific societies) coincides with that of the discourse paradigm which constitutes the regulatory system then this may be evidence of how dominant attitudes of the scientific community on GM food and crops are themselves generated by dominant cultural and political influences. We can, in turn, study dominant social, cultural and political influences on GM food and crops by looking at evidence such as opinion survey research, as well as other types of evidence such as the number and type of responses made to different official consultations.

In analysing what scientists do we must not put them on a pedestal and pretend there is something about science that insulates it and makes is fundamentally different from the rest of culture in society. Collins and

Pinch (2000: 142–143) say that it is a mistake to argue that because ordinary citizens do not know about the content of science that they cannot or do not wish to deal with conflicts of opinions among scientists or the practical value of scientific advice. 'Scientists are neither gods or charlatans . . . The expertise that we need to deal with them is the well-developed expertise of everyday life; it is what we use to deal with the plumbers and the rest' (2000: 143). Collins and Yearley advocate an approach which recognises that sciences are at 'one with our other cultural endeavours without making it necessary to deny that scientists have more skill, experience, and wisdom than others in the matters they deal with' (Collins and Yearley 1992: 308–309).

This attitude forms part of the so-called 'sociology of scientific knowledge' (SSK) strand which argues that 'scientific knowledge has to be seen, not as the transparent representation of nature, but rather as knowledge relative to a particular culture, with this relativity specified through a sociological concept of interest' (Pickering 1992: 5). So, according to SSK, scientists agree paradigms because of their common (perceived) interests, often meaning common pressures for progress in particular technical directions; pressures which emerge from society rather than being internal to science itself.

All these discussions can then lead onto a discussion of the political and social origins of the phenomena which have emerged as objects of scientific discussion and experimentation. We need, therefore, to turn to the theoretical basis for analysing how political outcomes occur.

Science, politics and society

In this second part of the chapter I want to develop an approach suitable for analysis of the wider social and political forces which act as the drivers of the discourse-paradigms. This approach also needs to facilitate a comparison between the politics of GM food and crops in our different case studies. In part, simply comparing the nature of the discourse paradigms that characterise the regulatory approaches to GM food and crops in the USA, the UK and the EU achieves these objectives. However we also need to study the actions of actors such as biotechnology companies, environmental groups, governmental institutions and we also need to study the attitudes of the wider public. I also want to deal with analysis of the role of the lay public in the regulatory debates, and also to prepare the groundwork for some conclusions concerning how public participation in such issues might be best achieved. The rest of this chapter deals with these issues.

Politics and power

I want to extend my use of Foucauldian discourse approaches (discussed in the previous chapter) to the analysis of how political outcomes occur and the notion of political power. One reason why I want to expand on this particular theoretical template is to emphasise how, and why, I am not in the business of saying that one or other agents is wholly responsible for this or that outcome. Similarly I am not in the business of determining that outcomes occurred and that attitudes were changed by specific events whose interpretation is unproblematic. It is convenient for people with partisan opinions, especially in the GM food and crops debate, to heap sole 'blame' on particular interests for particular outcomes. Alternatively, it is suggested that particular events, for instance the BSE crisis in the UK, are responsible for European hostility to GM food and crops. I think that a good understanding of the politics and sociology of policy outcomes demands a less simplistic account. Flowing from Foucault's de-centred notion of power is the idea that when one does political analysis one should be wary of attributing causation to specific factors and attributing power to specific agents to achieve certain objectives. For example, let us examine a statement made in a chapter about international political economy of biotechnology:

> Domestic pressure by environmentalists and the European Commission's desire to strengthen its role in maintaining food safety in Europe make it unlikely that the EU will simply give in to corporate lobbyists from both sides of the Atlantic.
>
> (Falkner 2000: 146)

Here the writer attributes power to pressure by environmentalists and the European Commission, neither of whom, (maybe for different reasons), will give way to the 'corporate lobbyists' (biotechnology companies). If this is to be an explanation of political outcomes we should try and discover how it is that the environmentalists have more power to influence European governments than environmentalists have power to influence the US government. No doubt the 'green' response would be to say that the corporations have more power in the USA and so on. However, corporations seem to get their way on other issues in the EU, so why not this one? Moreover, with regard to the above text's statement about the desire on the part of the European Commission to 'strengthen its role in food safety' we should ask why such a desire (assuming it exists) means that the Commission will be unsympathetic to pleas for authorisation of imports of GM food from the USA and cultivation of GM food in the EU. Needless to say, the author quoted above does not answer these questions.

I hope that the analysis of how political outcomes occur will be more rigorous than merely repeating popular pro- or anti-GM food/crops narratives which involve either globalising corporate interests or scaremongering Luddite environmentalists as the villains. Much political persuasion seems to depend on the attribution of 'conspiracies' on the part of one's opponents. If you can tie your opponents in with negative anti-democratic activities, then in today's world of dominant liberal democratic discourse, they are cast as the bad guys.

Hence the anti-GM food activists will imply that farmers have been forced to adopt GM crop technology through false propaganda about the non-existent financial benefits of shifting to GM cultivation. On the other hand the pro-biotechnology interests sometimes tell US audiences that, on the one hand the Europeans are 'protecting' their own technologically inefficient food producers by keeping out GM food imports, and also that the European consumers have been scared off buying GM food because of irresponsible scaremongering of environmental groups.

As we shall see, these narratives are matters of interpretation. I do not claim to escape from my own narrative into some elusive objectivity. However, I deploy an analytical narrative to the extent that I am trying to analyse the role of the dominant narratives in the GM food and crops debate without consciously acting as a propagandist for one or other of them. Foucault's notion of power is a useful way of performing analysis of how different discourses gain dominance in differing circumstances because he was wary of attributing causation to particular agents and of interpreting events in unidirectional ways. Foucault discussed political and social outcomes as occurring in the context of a seamless web of power relationships. Once society had moved beyond autocracy, a state which rarely exists today outside a few dictatorships, even Presidents and Prime Ministers are strictly curtailed in their room for political manoeuvre. They are curtailed, not only formally, by the demands of maintaining party majorities and the need to win sufficient support to implement policies but also discursively in terms of the need to concentrate on trying to say the right things in the right way according to the dominant publically accepted truths of the day. To me stories about 'conspiracies' are tropes, expedient tools, to brand opponents as evil-doers. Whether one believes in a conspiracy or not in this case is to make more of a statement about what you want to believe than a statement of absolute truth about who has power.

We must also look at how events are interpreted rather than merely assuming that events can only have a single set of political consequences. The assumption that changes in attitudes or other events are caused by events whose meaning is taken for granted is a positivist error, just as is the attribution of cause to specific agents. Rather, an interpretivist account will seek, where appropriate, to investigate how events came to be interpreted in a certain way. For example, a storyline about a food problem may have quite a different impact on a population (maybe the EU?) that is more sensitive to food style/health issues than one which is less sensitive to food style/health issues (maybe the USA?).

In summary, I am not out to look for conspirators, or imply that one or other set of actors has forced practices on another set of actors. In addition I am not out to establish events as the sole causes of outcomes. On the other hand it may be useful to deploy a pragmatic, rather than dogmatic, version of Foucault's analytical techniques. In other words we need to examine the role of agents and also of key events, even though the precise causal relationship between, on the one hand the effects of agents and changing events and, on the other hand, outcomes are impossible to determine.

A pragmatic approach to Foucault

The tactics employed by agents can be associated with effects. It is just that actions by agents in one context may not have the same effect as the same actions by agents in a different context. For example, one plausible hypothesis is that environmental campaigning against GM food and crops may have had much more resonance with consumers in the EU than the USA in the 1998–2000 period.

I doubt whether American farmers have always adopted GM food technology simply because they were tricked or forced into adopting a technology that never has any commercial farming benefits (although there are many doubts as to whether current versions of Bt corn are much use! – hence its low take-up). On the other hand it is apparently the case that without the activities of biotechnology companies farmers would not (could not) have adopted the technology. Decisions about what sort of GM crops to offer to farmers and the ways in which biotechnology companies have attempted to validate their claims about environmental improvements has vitally affected outcomes in differing contexts. Many argue, for example, that GM crops will fail to feed the world's poor simply because biotechnology companies will not develop GM crops for people who have little money to buy them.

While it does seem that European consumers have not been put off buying GM food simply because of scare stories spread by environmental groups (consumers were pretty scared already), it seems equally valid that they needed to be organised before such fears could result in effective consumer boycotts. Without the active organisation of consumer protests by environmental groups and other anti-GM groups the emptying of GM food from supermarket shelves would certainly not have been so rapid, and may not even have occurred at all. Equally, without the active political intervention by environmental groups, it seems unlikely that there would have been a moratorium on GM crop and food authorisations in the EU. Further clarification of such issues will be made in Chapters 3, 4 and 5.

Critics of Foucault (and even some of his advocates) argue that discourse analysis is implicitly unable to compare the accuracy of rival narratives. The very epistemology (theory of truth) of Foucauldian discourse theory assumes (for the purpose of analysis) that you can have no knowledge of reality that is external to discourse, or at least external to a particular discourse. You cannot adjudicate on the comparative veracity of differing discourses. Yet there are some practical limits to this which make it unwise, and indeed, unnecessary, to rule out all attempts to adjudicate between the truth of different discourses. In practice all human societies have some agreement about reality simply through having human senses, human physiology and human emotions (Toke 2000a: 162–166; Bevir 1999). Admittedly a fish-like alien intelligence would have a rather different ontology to our own, since they drown on land while we drown in the sea, but this is hardly relevant to a discussion of GM food and crops!

There is a debate about whether humans can conceptualise the interests of non-human species, and therefore pursue an ecocentric concern for the preservation of nature independent of human interests (Eckersley 1992: 106–117). Habermas disagrees with the notion that we can do this and says we can only base our ecological concerns on 'aesthetic experience and feelings' (Habermas 1982: 243–244). However, this debate need not concern us here, because we are dealing with an issue that is not only founded on very human notions of reality but one that is founded in late twentieth and early twenty-first century notions of reality inside the USA, the UK and the EU. There are many common, agreed, notions of reality here. Part of my aim is to look for the differing conceptions of reality that have led to differing political outcomes in these cases. This is in keeping with a conventional discourse approach which seeks to describe different storylines. However, if we can talk also about agreement of various aspects of reality, we can also look farther than discourse and take into account techniques such as public opinion surveys or responses to public consultations organised by governments. These can be used, for example, to discuss which storylines seem to have greater (or alternatively, less) relevance to different publics. We can also look for associations between differing events (or other pieces of evidence) in order to try to assess differing explanations of how events occurred. Now we cannot determine cause by such procedures, but we can at least begin to distinguish between those explanations that are relatively plausible and those which are not.

However, while interpretivist approaches do not rule out behaviouralist techniques, for example the use of opinion surveys, interpretivism does suggest that the results of such studies must be used carefully. First, results of opinion surveys are contingent on the questions asked. Second, one must be very wary of, indeed generally disposed against, attempts to compare opinion surveys which use different techniques and questions. Third, the answers themselves need to be interpreted. If people give a reason for a particular action one must be wary of ascribing this as a cause since you do not know whether something else in the context has made the reason given appear particularly important. I make this point because of the argument over the extent to which the BSE crisis is said to have influenced opinion to swing against GM food in Europe. People may cite BSE as a reason for objecting to GM food, but if there were pre-existing

European sensitivities to genetic modification of food and generally to food safety and food quality is it reasonable to simply ascribe European hostility towards GM food to the BSE crisis? I shall discuss these issues further in Chapters 3 and 5.

Discourse theory opposes the notion of establishing causes for events since no single statement or set of statements can establish the whole truth about cause. Yet there is great pressure, on practical grounds, to attribute cause in order to focus attention on how we can improve arrangements for the future. Philosophically, it may be correct to say that a given train accident was caused as much by the invention of railways or the fact that the driver decided to set off at all. Unfortunately this will not help us improve safety on the railways. Similarly, in political and social analysis, if we want to change things for the better, or at least engage in analysis that can be summarised in a more popular and therefore under-standable form, then prudent analysis of how outcomes have transpired and the role of specific agents and coalitions can be a useful activity. In order to do this we need to examine the contexts, including the changing dominant discourses and power relationships that exist.

In saying that it is often useful to discuss the role of specific agents is to make an implicit criticism of Foucault, who seemed relatively uninterested in the influence of agents. He argued, in most of his work, that the individual subject cannot escape the dominant set of discourses which dictate what his/her actions would be. I agree with the position of Bevir (2002) who argues:

> I do not believe that postfoundationalism [meaning that social truths are related to context] requires us to take the overwhelmingly negative view of the subject we find in most of Foucault's work. To me postfoundationalism undermines the idea of autonomy – that the subject can adopt beliefs or perform actions uninfluenced by society – but it leaves intact the idea of agency – that the subject can reflect on her beliefs, modify them, and so perform novel actions for reasons of their own ... However we still need some way of explaining why traditions change.
>
> (Meyet, interview with Bevir, 2002)

Bevir also criticises Foucault's approach by saying that it can only work as a neutral device to analyse power and it cannot say that one power system is less just than others. I agree with this criticism, but I disagree with Bevir's conclusion that this drains the importance, for political analysis, of Foucault's notion of power. As I have argued earlier in this Chapter, let us use Foucault's notions of discourse and power as a 'neutral' mode of analysis, which, apart from anything else, can be used to criticise conspiracy theories and look for power relations which condition and constrain the individual subject (agent). This theory of power still acts to deprive dominant (scientific) groups of sole claims to the control of truth, even if this notion of power does not sanctify other (scientific) authorities as possessors of truths either.

The notion that the power of agents is constrained by wider power relationships is shared by Marxists and neo-Marxist critical realists. They will seek, among other things, to analyse how economic relationships constrain agents. Foucauldian thought, in contrast, dismisses any notion that economic structures can be defined externally to discourse. However, this contrast between a Foucauldian approach and a critical realist approach is not so clear in the case of some critical realists, such as Hay (2002), who recognise the impossibility of specifying an exact relationship between the influence of such economic structures and how agents act.

Hay argues:

> Political actors inhabit complex and densely institutional environments that favour or privilege certain strategies over others. Yet such actors do not appropriate these contexts directly, blessed with a perfect knowledge of the contours of the terrain. Rather their ability and capacity to act strategically [that is to act to influence what others do] is mediated and filtered through perceptions (and indeed misperceptions) of the context they inhabit.
>
> (Hay 2002: 57)

This opens up considerable space for the influence of ideas, and thus the potential use of discourse analysis to chart the course and outcome of battles over such ideas (Fischer 2003: 21–47). Hence critical realists may regard discourse analysis as a potentially important means of analysing changing ideological paradigms, even though they diverge from Foucault in seeing that there is some relationship between separable economic and discursive structures. Hay goes on to argue later that:

> [I]t is, frankly, implausible to suppose either that actors have complete information of the context in which they find themselves or that their behaviour is rendered entirely predictable by the (presumably transparent) material interests they hold in a given context. Yet qualify either of these assumptions and an independent role for ideas is immediately opened within political analysis.
>
> (Hay 2002: 257–258)

As we shall see in Chapter 4, it is plausible to argue that perceived material self-interest of US farmers who are growing GM crops produces a powerful lobby opposing labelling of GM food in the USA. Hence it would be foolish to omit reference to economic factors. Yet at the same time one must remember that these are perceived interests, for instance that labelling GM foods will render them less saleable and therefore hit farmers financially. Alternative notions of what constitutes farmers' material interests may exist. It may be argued that segregating and labelling US GM produce may gain access to world markets which are increasingly demanding labelling of GM food and crops, hence altering such perceptions of US farmers' self interest. As we shall see later, it is, apparently, plausible for North American wheat farmers to argue against the introduction of GM wheat because, while it may lower their own costs, it would effectively bar their products from GM-averse European markets. The battle for supremacy of the interpretations of such self interests is one that can profitably be studied through the realm of discourse analysis. Here lies an area of agreement between Foucauldian analysts and critical realists such as Hay who place emphasis on ideational influences. Collaboration between Foucauldian and critical realist approaches (Toke and Marsh 2003) may proceed on the basis of such understandings.

However, a Foucauldian approach, which is of central influence to this study, would argue that what passes for material interests are, at the end of the day, mere interpretations. Material structures are 'real' only inasmuch as there is an agreed notion of the nature of a material interest. Even the value of money and material assets depends on belief systems, as investors in the shares quoted on the stock exchanges will be keenly aware.

We need now to discuss the tools of analysis that can operationalise this approach to the question of power and the role of ideas. The notion of discourse analysis has been explored earlier in this chapter, and much of the technique can be transferred from looking at regulation about legal and scientific issues to looking at how political actors form coalitions.

Ideas, networks and coalitions

Hajer (1995) has linked discourse and political alliances through the concept of discourse coalitions. He argues that:

> The coalitions are unconventional in the sense that the actors have not necessarily met, let alone that they follow a carefully laid out and agreed upon strategy. What unites these coalitions and what gives them their political power is the fact that the actors unite around specific story-lines that they employ.
>
> (Hajer 1995: 13)

Hajer's method has the advantage of linking discourses to agents. However, he emphasises that the starting point of such analysis is the discourse rather than the interests of the agents. 'Interests cannot be taken as given a priori but are constituted through discourse' (Hajer 1995: 51) Hajer distinguishes his approach from that of the widely used advocacy coalition framework (ACF) theory, which talks of advocacy coalitions which involve a 'variety of actors' that have 'a set of basic values, causal assumptions, and problem perceptions – and who show a nontrivial degree of coordinated activity over a period of time' (Jenkins-Smith and Sabatier 1993: 25). Hajer argues in counterdistinction to ACF theory that:

> I aim to position myself . . . against studies that see social constructs as a function of the interests of a group of actors. In that case language is seen as a means and it is assumed that actors use language purely as a passive set of tools. In actual fact there is much more interaction between the linguistic structures and the formation of preferences . . . language is seen as an integral part of reality, as a specific communicative practice which influences the perception of interests and preference.
>
> (Hajer 1995: 59)

A detailed comparison of Hajer's notions and ACF theory is made by Fischer (2003: 94–116). It is certainly difficult to see how there is a clear set of 'basic values', as demanded by ACF theory in the case of UK GM food and crop policy between, on the one hand wildlife protection campaigners and, on the other hand, consumers who fear health risks (in the UK case). So how do such groups manage to collaborate? We need to discuss how language is used to construct interests and alliances between groups with sometimes disparate conceptions of self-interest. We also need to discuss how events themselves are constructed.

ACF theory makes the implicit positivist assumption that events have one interpretation (Jenkins-Smith and Sabatier 1993: 29). As I explained earlier, I employ an interpretivist (or social constructionist) approach which examines the way events and issues are constructed in some ways rather than others (Toke 2000b: 839; Hannigan 1995). Indeed, the centrality of events themselves in explanations of how ideas change is a major potential weakness in ACF theory. As I demonstrate in Chapter 5, by examining the role of the BSE crisis in the development of EU GM food and crop policy, it is all too easy to oversimplify the impact of events. BSE cannot be accepted as the sole, or even as the principal, cause of European hostility to GM food and crops. This is because considerable differences in approach to the issue existed in the EU compared to the USA before the BSE crisis erupted in the UK (on 20 March 1996).

One way of conceptualising how different groups come to relate issues to their own pre-conceived interests is the notion of 'issue framing'. This was conceived by Goffman (1975) and I deploy the concept in Chapter 3 to study how different groups mobilised in opposition to GM food and crops in the UK. Erving Goffman said, in connection with newspaper reports, that 'Our understanding of the world precedes these stories, determining which ones reporters will select and how the ones that are selected are told' (Goffman 1975: 15). As we shall see in Chapter 3, various groups in the UK became critical of GM food and crops, but did so for rather different reasons that were related to their own notions of self-interest.

Hajer bases much of his analysis around the study of storylines to which groups align themselves. He discusses how metaphors can be used to invoke storylines (Hajer 1995: 61–63). For example, in the GM food and crop case (in the USA) the monarch butterfly is invoked to represent threats to wildlife from genetic engineering. On the other hand 'golden rice' is invoked to represent technological gains to be derived from GM crop technology.

We can talk about networks of groups that relate to these storylines. Such networks are by their nature impermanent, relying on sometimes changing storylines for their cohesion. Sometimes there will be more than one type of alliance behind different storylines on different aspects of the same issue. For example, in the USA the grocery manufacturers tend to ally with farmers and biotech companies in support of GM food and crops, but on the issue of plant-based pharmaceutical products their sceptical approach has a lot in common with actors such as the Union of Concerned Scientists who are usually associated with positions that are critical of GM crops. Network analysis is important to my study, but it is a tool which deals with a single layer of ideas, that of a discourse or storyline to which an alliance of groups are committed, rather than imagining a set of 'real' group interests which underpin the ideas.

So, we can trace how networks of perceived interests interact and how their belief systems and practices develop, but what about explaining how political outcomes emerge from the governmental process? We need a tool of analysis that investigates governance from the inside.

One variant of network analysis, that is the policy network approach expounded by Marsh and Rhodes (1992), sets out to explain outcomes by reference to the type of policy network which operates at the level of government decision making. This has been deployed by Toke and Marsh (2003) to chart changes in GM food and crop policy making in the UK. The type of governmental decisions will be related to the groups that are included in closed policy communities that make decisions, and if the membership of such policy communities changes, then so will the decisions that emerge from them. The membership of the policy community will be related to the dominant discourse which underpins that network (Toke 2000b).

In the case of the EU where key networks of NGOs are focused at the national level, it seems more appropriate to focus on how the issues are dealt with through the different institutions. Hence we can study the activities and influence of the European Parliament, the European Commission and the Council of Ministers, and the scientific committees that are associated with the GM food and crops issue.

Discourse and institutions

The mention of the need to study of European institutions leads me onto consideration of how my discourse approach can be applied to the study of institutions in general analysis. There are various self-styled notions of 'institutionalism' which use institutions, defined as rules and practices (Pierson 1993), as the focus of analysis.

According to Hall and Taylor (1996) there are three types of institutionalism. So-called 'rational choice' institutionalists tend to look at how institutions affects political behaviour by altering the incentive structures faced by actors engaged in rational pursuit of their own self interests. On the other hand 'sociological' institutionalists look at issues such as the very constitution of self interest and how behaviour is shaped by dominant norms and belief systems which form the very structures that constitute institutions. Hall and Taylor argue, in their 'historical institutionalist' approach, that there is room for a dialogue between the cultural approach (associated with discourse theory) and calculus approach (associated with rational choice theory).

There are those who suggest that such a project is doomed from the start. Hay and Wincott (1998) argue (against Hall and Taylor), that rational choice and sociological (cultural) approaches are based on 'mutually incompatible premises or 'social ontologies' (1998: 951). I tend to side with Hall and Taylor in this dispute in the sense that while Hay and Wincott are correct in identifying different methodologies, I feel that the 'cultural' and 'calculus' methodologies can both develop explanations which the other cannot. Many, if not most, people would agree that art and mathematics are 'incommensurable' in the terms described by Hay and Wincott. Yet both disciplines are used side by side in the process of designing buildings. Indeed their complementary use is essential to successful building projects. Feyerabend, the philosopher of science who emphasised the incommensurability of differing scientific theories, was also keen to emphasise how a 'proliferation' of theories should be deployed to increase our understanding.

Historical institutionalism is said to be especially relevant to comparative studies of policy/political outcomes. Many of the most cited historical institutionalist accounts of comparative politics have dealt with how constitutional rules produce different outcomes in different countries. For example, the case studies in Steinmo *et al.* (1993) concern the effects of different party systems, the number and disposition of 'veto' or 'entry' points for policies into government, electoral systems and other constitutional differences.

However, a more traditional use of institutionalism is simply to analyse the processes, perceived interests, resources, policy standpoints of particular institutions as organisations (both within and outside government) that are closely involved in the issue. Interestingly, the shift from an emphasis on organisations to one based in rules is said by one analyst to be a feature of 'new' as opposed to 'old' institutionalism (Lowndes 2002).

Scott (1995) identifies three types of structure making up institutions. These are regulative, normative and cognitive. Regulative structures are the rules that alter actors' incentive structures; structures which are the central focus of rational choice institutionalists. Normative structures prescribe actors' interests in terms of following roles or norms. Cognitive structures refer to common world views of goals and reality. Cognitive structures are studied through discourse analysis, and hence cognitive structures are much more of the focus of analysis (in this study) compared to the other two types of structure. An alternative approach is to conflate the notions of 'cognitive' and 'normative' structures into the concept of 'the logic of appropriateness' as a guide to action by actors in particular political settings (March and Olsen 1989: 159–161).

My own preoccupation with belief systems and discourse would place this study, in Hall and Taylor's terms (1996) much more in the realm of sociological institutionalism compared to rational choice institutionalism. I focus mainly on the study of belief systems (the cognitive rules) since they are the key to understanding the different approaches to science in the EU and the USA. Differences between, and changes in, the GM crops and food regulatory systems are based on different ideas. The only indisputable portal of access to these ideas is the study of discourse. Hence it is this focus, on cognitive structures, that is the strongest focus in terms of the way Scott (1995) would discuss institutionalist theory. However, a distinction needs to be made between two types of cultural approach. First, sociological institutionalists, such as DiMaggio (1991), who, influenced by Bourdieu (1993: 72), look for generalisable theories of cultural behaviour and, second, a discourse analytical approach, used in this work, which assumes that analysis is specific to a particular case study.

I study the discourses which together make up regulatory institutions, that is the laws, rules and the scientific advisory bodies dealing with agricultural biotechnology. In addition, I focus attention on path dependence through attention to pre-existing cultural attitudes and policy decisions on the role of regulation and industry, the level of trust in environmental regulation and regulators, nature conservation and food. As such, I do pay attention to the notion of path dependence, a hallmark of 'historical' institutionalist accounts. However, my emphasis is the influence of traditions on the evolution of regulatory discourses and the nature of the networks which are associated with dominant discourses and storylines.

As discussed earlier, we can also study the differences between the policy networks in different countries, the nature of the discourses that underpin those networks and the nature of the organisations that make up these policy networks. For example, there is no equivalent in the USA of English Nature, the statutory body which has an interventionist role in UK agriculture (and which demanded special field trials of GM crops).

They have become closely involved in the scientific evaluation of GM crops. This reflects the British environmental tradition as seeing environmental protection as being concerned with farmland nature conservation issues. In the USA, the Department of Agriculture (USDA), the Environmental Protection Agency (EPA) and the Food and Drug Administration (FDA) have dealt with environmental and health evaluation of GM food and crops. However, biotechnology companies are widely (and by anti-GM activists, notoriously) associated with taking the lead in environmental aspects of US GM crop policy and they are accused of having close personnel links with the USDA, EPA and FDA (Anderson 1999: 93–97). This may reflect a dominant discourse which privileges industrial interests in agricultural biotechnology more prominently compared to many European countries.

I utilise, at times, the notion of 'path-dependence' which is so emphasised by historical institutionalist accounts, and I also pay some attention to the possible influence of rules as incentive structures, the influence of norms and the specific contribution of agents. However, the core of my approach concerns how belief systems – as accessed through study of discourse – differ in the case studies and how different networks of interest groups associated with different storylines have dealt with the GM food and crops. Rather than adopting the ascription 'historical institutionalism', (or even the narrower term 'sociological institutionalism'), I would prefer to say that the pragmatic 'discourse-analytical' approach I use employs many of the techniques associated with analysts working from an institutionalist background. After all, however institutions are defined – whether broadly as rules and practices, or narrowly as formally constituted organisations – we can study these phenomena through discourses. As Howarth puts it:

> (In discourse analysis) Institutions are conceptualized as *sedimented* discourses. In other words, they are discourses which, as a result of political and social practices, have become relatively permanent and durable. In this sense there are no *qualitative* distinctions between discourses, only differences in their degree of stability. This means that relatively fixed discursive formations such as bureaucracies, states and political parties are legitimate objects of discourse analysis.
>
> (Howarth 1995: 132)

There is a self-contained discourse on the theory of EU institutions and European integration which deals prominently in some themes that could be relevant to this study. I want to have a brief look at some of these themes.

EU theories

Perhaps the most fundamental (and often abused) theory of European integration is the notion of 'functionalism'. This is both a normative wish by Monnet, one of the founders of the European Union, and an academic theory that European integration proceeds by harmonisation of regulatory activities in technical fields which then produce a political 'spillover' effect which generates closer integration at a political level. As one analyst puts it:

> The founding fathers of the EEC, notably Jean Monnet and Robert Schuman, firmly believed that the cause of political integration would be best served if the Commission focused its energies on building transnational cooperation in the areas of 'low politics' in the hope that enthusiasm for European solutions would eventually 'spill over' into more sensitive areas such as the economy and foreign and defence policy.
>
> (Jordan 2002: 5)

So the first question is whether there is evidence of such a trend in this case study. A second, perhaps less controversial, notion has been elaborated by Majone (1992) about pressure for regulatory harmonisation from industrial groups. He argues that, as in the USA, multinational, export oriented industries press for EU wide legislation in order to avoid inconsistent and progressively stricter regulations in individual EU states. Majone cites the case of US car emission regulations (Majone 1992: 306). To what extent can one extend Majone's analysis to the GM food and crops issue?

Third is the idea, propounded by analysts such as Cram (1997: 176), that the European Commission has succeeded in overcoming the objection of member states to its designs for measures of harmonisation of technical matters. Regulation of GM food and crops is certainly a technical issue in many regards, so we might ask what relevance such ideas have in this case study.

Fourth is the question of the relationship between the Commission, the Council of Ministers and the European Parliament on environmental policy issues. Textbooks on EU environmental policy regard as almost axiomatic the proposition that the European Parliament is the natural ally of the Commission on environmental matters. The Parliament helps the Commission to cajole reluctant member states to agree to rigorous, harmonised, environmental regulations (Judge 2002: 128). One example, in this case the example of GM food and crop policy, cannot be used either to confirm or discount this widely accepted nostrum, but it may be instructive to see if there are any differences from 'normal' practice in this case, and if so, consider what is different about this issue.

Related to this is the fifth issue, that of the power of the European Parliament. It has become accepted wisdom that since the Single European Act (which took effect in 1987) the European Parliament has become a significant influence on EU legislation. The car-exhaust legislation is cited in support of this notion (Weale 2002: 339). Is the European Parliament a significant influence on GM food and crop legislation? If so, could this issue become a more up-to-date (compared to car-exhaust regulations) exemplar of Parliamentary influence?

These five issues of theory have been selected in order to compare processes and outcomes in the GM crops and food issue to established theory of European integration. However, there are other modes of analysis which follow from the historical institutionalist approach I described earlier and which might be specifically applied to the EU in order to help explain outcomes. In particular there is analysis of the regulative, normative and cognitive rules which Scott (1995) has described as making up institutions. I have already explained how my central focus is on the cognitive structures, or belief systems as analysed through discourse. However, I also look at the influence of norms and of incentive structures, and these latter rules may be important in the case of the EU.

The different institutions that make up the EU, the Commission, the member states and the European Parliament, all have their own norms which influence their actions. I discuss some important norms in Chapter 5, such as how the Commission deals with situations where it cannot find consensus policies. I also look at the extent to which regulatory structures have influenced outcomes. For example, differences between the two-level nation/state scientific regulatory committee set up in the EU and the single federal level structure in the USA may, relative to the USA, inhibit the ability of the EU Commission to carry out its legislative obligations to promote biotechnology products that it has authorised.

I have tried, in this section on EU theory, to talk about tools of analysis that are specific to the EU. It would be repetitious to talk about the tools of analysis such as discourse analysis and network analysis which I have already described. Nevertheless I apply these tools to the EU as well as the other two case studies.

As I have mentioned earlier, my discourse – analytic approach is eclectic. I choose my tools pragmatically so as to be useful in analysing this GM food and crop comparative study. I also think it useful to study the influence of public attitudes by use of existing opinion research evidence. It would be a mistake, in this case, to imagine that any one of the networks of interest groups and the discourses that underpin them represented public attitudes. Yet public attitudes are crucial in this issue. Consumers and voters will decide whether they want to buy GM food and whether they are willing to pressure their political representatives in support of policies such as labelling GM food products. How do we analyse the role of the public?

Role of the public

Just as I have argued earlier that positivist outlooks ignore the process of how events and belief systems are socially constructed, so analysis that relies solely on discourse can miss important evidence about the nature of belief systems and associations between belief systems, political behaviour and political outcomes. Quantitative empirical analysis of people's behaviour, in this case analysis of the general public's behaviour, can increase understanding of such linkages and consequently how political outcomes occur. This technique is known as behaviouralism. There are some questions that discourse analysis, on its own, is unable to answer. In particular how is it that some storylines are more resonant with the public than others? We may not be able to ascribe specific cause and effect relationships, but collection of evidence from, for example, carefully researched opinion surveys, can help us separate out the possible from the unlikely. Such behaviouralist techniques will not, like discourse analysis, provide all the answers, but together with discourse analysis we can build a clearer picture of how political outcomes may occur in this case of GM food and crops.

Many analysts subscribe to the view that the public's attitudes to science and trust in experts is a crucial aspect of this issue which has to be analysed in its own terms. It is to the public attitudes to science, risk and trust in experts that I now turn.

Science and society

Irwin (1995: 14) has characterised the traditional attitude of the scientific establishment as one that sees the role of the public as being no more than an object to be educated out of its irrationality and ignorance. He calls this a 'science centred' view. Yet such a view is clearly out of date. Official science (at least in the UK and the EU more generally) faces a growing credibility problem as an increasingly educated and aware public sees instances when official science has been forced to admit that the opinions held by members of the lay public turned out to be true after all. The issue of transfer of BSE to humans is one that is very relevant to the context of distrust of GM food technology, as we shall see in Chapters 3 and 5. However, there are plenty of other instances where scientists have, to their costs, ignored the opinions of lay-observers.

Wynne (1996) tells the story of how post-Chernobyl analysis of radioactive contamination of Cumbrian (UK) farm environments was hamstrung by a failure and refusal of scientists to listen to farmers' knowledge of local soil conditions and sheep behaviour. Regulators have a habit of basing their risk assessments on the assumption of ideal sets of farming practices. For example, the UK's Advisory Committee on Pesticides (ACP) rejected demands by the National Union of Agricultural Workers that a herbicide that had already been banned in the USA and Canada should be banned in the UK. The ACP maintained that its own tests demonstrated that there could be no significant risk of the medical problems, including birth defects, that were said to be associated with the herbicide. Yet the Union pointed out that in practice the manner in which the herbicide was handled was very different from the models used by the regulators for risk assessment purposes (Irwin 1995: 17–21). There are plenty of other examples featuring knowledge about indigenous peoples' knowledge of local flora and fauna being superior to that of mainstream scientists (Edwards 1998).

One recently (at the time of writing) publicised example of the dismissal of lay knowledge about the natural concerns the behaviour of waves in the seas (BBC 2002). Until the mid 1980s the so-called linear model of waves was dominant among physicists. Recurrent tales from mariners of 'freak waves', waves of 30 metres in height which nearly sunk their vessels, were dismissed by the physicists. However, in 1985, a wave measuring device installed on an oil rig in the North Sea near Britain did indeed record a 'freak wave'. The scientists began to look at other, dynamic, models of wave behaviour, and now there has been a paradigm shift in theories of wave behaviour towards acceptance of the existence of 'freak waves'.

Given the existence of examples like this, many feel uneasy about the rapidity with which official scientific spokespersons dismiss public fears about GM. That, of course, does not necessarily mean that the public fears are right and the official scientists are wrong, but it does not reflect well on scientists who are arrogant in their condemnation of public ignorance about scientific matters. In democracies the attitudes of the public are vitally important, especially if their fears are legitimised by well organised interest groups, in particular environmental groups, whose perceived interests overlap with the public's scepticism towards GM food technology. Beck's sociological analysis of risk interprets lay questioning of the infallibility of official scientific judgements as being evident of the shift from 'primary scientisation' to 'reflexive scientisation' (Beck 1998: 158–163). This means a shift from seeing science as the midwife of social and economic development towards seeing it also as the creator of risks. However the central focus of Beck's interpretation of the relationship of scientists to lay people is on how the new technological threats are seen as being 'scientifically constituted' and how 'lay protest' is dependent on 'counter-scientific mediations' (Beck 1998: 162). This deprives non-scientist opinion of any substantial role in opinion formation. It reduces them to spectators who signify their support for one or other contesting groups of scientists. But the role of lay-publics is not like that.

We ought to be wary of just lumping lay public and environmentalist concerns about technology together, as if lay publics simply decide to support environmentalist as opposed to official science. As we shall see in later chapters, the public tends to adopt different priorities on the GM food and crops issue compared to the environmental groups, even though they will back general demands for moratoria on crop commercialisation and food sale licenses. There is a tendency to spend a lot of time analysing the environmental groups' activities and their attitudes to risk. We need also to study public assessments of risk which differ qualitatively to those of both official science and the environmental groups.

If the epistemology (rules of knowledge creation) of risk for natural sciences are concerned with positivist assessments of narrow issues determined by officially appointed scientists, then the epistemology of lay knowledge of risk is concerned with a broader and implicitly interpretive approach that openly includes consideration of issues that are beyond the remit of official science. A study of the attitudes to risk held by residents close to a petrochemical plant in North West England found that 'The point, of course, about a citizen-based perspective (to science and risk) is that issues such as pollution and major hazard threat will inevitably blend in with the wider background of other local concerns' such as the issue of jobs (Irwin 1995: 93–94). In an exercise to discuss the siting of landfills in Aargau in Switzerland the recommendations of the canton's building department was rather different from those of the citizens' panels. 'The main reason for this difference was the high importance that the officials from the building department assigned to geological stability whereas the citizens included social and aesthetic criteria in their preference judgement' (Dienel and Renn 1995).

One explanation of these differences is simply that the implicit interests and therefore the risk/benefit assumptions of official scientific regulators and the lay public are different. For example, for good or ill, scientists who sit on regulatory committees may implicitly take into account dominant discourses on the gains for economic efficiency that will result from harnessing new technology. The scientists, who are confident in their own risk-assessment models, may feel that the risks to health and environment are of small or of marginal concern in comparison to the economic gains. On the other hand if the public do not see any obvious gains to them, as individuals, in adopting a new technology, but feel there are possible risks which are given legitimacy by a mixture of experience of being let down by official scientists and the opinions of dissident scientists operating maybe from different standpoints, then they may well react with scepticism towards the technology. Lay citizens have, in many ways, collected their own empirical evidence about the track record of 'official' science, and the hypothesis that says that they sometimes get it wrong seems to be reasonably well supported by evidence.

A further explanation of the different approaches to risk by lay publics and official scientists is simply that they mean different things by risk. Official science deals in quantitative data about the chances of fatality per year, but research suggests that ordinary citizens base risk judgements on various qualitative factors.

In theoretical terms, this underlies Beck's assertions that the risk management procedures that have been put into effect in the era of secondary scientisation have failed to solve the public's fears about technological risk. One research analyst, speaking from the point of view of a social psychologist, says of attempts to deal with 'the causes of the wide gaps that occur between 'expert' and lay evaluations of the risks of modern technology':

> The response of the technical community to these often profound disagreements has been to focus on the search for more accurate methods of measurement. However, probabilistic risk assessments do not inspire public confidence or close the gap. The lay person's risk perceptions are based on a much wider range of social-psychological, political and ethical considerations and are strongly influenced by the qualitative nature and attributes of the particular hazardous activity.
>
> (Lee 2000: 1)

Lee (2000) cites work by Slovic *et al.* (1980) on how experts and lay citizens rank risks very differently. For example, in a study of what different groups of citizens feel are the most serious technological risks, the League of Women Voters (in the Slovic study) and College students both put nuclear power as the number one risk. On the other hand, experts ranked nuclear power at only 20 in the order of risk and put motor vehicles as the number one risk. Yet again, non-nuclear electricity was rated by the lay citizens as 18th to 19th in the order of risk while the experts put non-nuclear electricity at number nine. Lee discusses Slovic's analysis of categorising the qualities of particular risk under factors such as 'dread', 'familiarity' and 'catastrophic potential'. Nuclear power scored highly negatively on all three factors. 'Dread' is said to involve lack of control and effects on innocent victims, familiarity is about whether the processes are 'well known to people' and 'catastrophic potential' is 'the number of people, that could possibly be killed in a single accident and the degree of personal exposure' (Lee 2000: 11) This type of analysis can help us understand how technical and lay risk assessments can vary, although it may not be so good at explaining differences in risk assessment between different sets of lay public.

This type of analysis has been applied to GM food. Frewer (1999) has focused on the 'control' aspect. She focuses on the fact that the early imports of GM into the EU were not labelled. She identifies this as something that prompted consumer concern over lack of control and mistrust of the food industry. While this explanation sounds elegant, it does not explain how in the USA, where there has been no labelling of GM food, there was much less consumer resistance to GM food compared to the EU. This emphasises a shortcoming of Slovic's explanation of the difference between expert and lay assessment of risks. It is good in general terms, but still leaves us the task of explaining how different lay publics have different assessments of risks for the same technology.

Citizens make up their minds in the context of a range of discourses about risk and benefits. Scientific information about risks will feature prominently in this. However, research suggests that the public chooses which conflicting stories to believe on the basis of indicators of trust attached to the source of the information. Studies on the trustworthiness of sources in the UK on food-related risks indicate that the lowest trust (among the sources recorded) is assigned to statements made by government ministers and Members of Parliament, followed by tabloid newspapers. Consumer organisations, quality newspapers, television broadcasts (especially documentaries), university scientists and medical doctors are more solidly trusted; medical doctors being the most trusted of all (Frewer *et al.* 1996: 481).

Conclusion

If we are trying to analyse how outcomes have transpired in a complex social, scientific and political issue such as GM food and crops then it should not be implausible that a number of tools are going to contribute. My prime focus of analysis is an investigation into the dominant discourses which underpin GM food and crop regulatory discourses in different regimes. I use the concept of discourse-paradigm to describe the regulatory frameworks. As the title implies, this borrows ideas both from Foucault's investigation of the human sciences and Kuhn's study of scientific development in the natural sciences.

The usefulness of this approach for this comparative study rests partly on two of its central aspects. First, the study of key statements to compare truth systems which legitimise often different types of knowledge. Second, the paradigm concept which focuses our attention on issues raised by Kuhn including the type of fact collecting activity associated with each discourse-paradigm and the nature of the scientific communities associated with each discourse-paradigm.

The concept of discourse-paradigm implies that social influences decide what constitutes the truth concerning issues such as whether GM plants are different to conventional plants. Social influence imbues the scientific community in a particular country with distinctive dominant interpretations of scientific uncertainty and notions of what types of knowledge need to be collected by the scientific regulatory system.

In using an 'interpretivist' type of analysis I avoid positivist errors. Such errors include the assumption that issues involving scientific uncertainty are, or can be, settled solely by scientists. Other types of error committed by positivists are the attribution of sole cause to specific agents, the attribution of sole cause to events and the assumption that particular events have only one plausible interpretation. If we are to understand the GM food and crops issue, at least in a comparative context, it is vital that we study the context in which agents act and, also, the ways that key events are constructed and interpreted.

My approach also involves utilising various other tools, such as network analysis, analysis of storylines and also use of opinion survey analysis to discuss the political and social underpinnings of the discourse-paradigm and explanations for differences between the discourse-paradigms which we are comparing. Discourse-paradigms are underpinned by dominant policy networks. We can analyse the nature of these dominant policy networks, and changes in them. Particular networks are held together by what Hajer calls 'storylines'. By focusing on the use of storylines we can trace changing alliances and their association with political outcomes. I place great importance on Foucault's de-centred conception of power. In the context of this study this means that we disavow conspiracy notions which blame particular groups for particular outcomes. We can also use non-discursive tools. For example, we can obtain at least some notion of the resonance of particular discourses with the public by using opinion poll and panel survey data.

My approach utilises various techniques that have been associated with types of institutionalist approach. However, I think the dominance of the discourse approach means it is most accurate to talk about different techniques being associated with a discourse-analytical approach. The comparison of institutional arrangements is studied principally through comparison of discourses on key and cultural influences concerning environmental regulation and food. This includes a particular focus on the key regulatory and scientific discourses in the different regimes, and a study of how the differences in discourse have emerged. I focus on organisations, how their traditions affect their current stance and how this interacts with changing contexts to produce responses to the issues involved in GM food and crops. Hence some history of relevant controversies such as (in the case of the UK) the BSE crisis will need to be considered.

It is now time to deploy these various tools of analysis. The first case study I examine is that of the UK. The UK is an interesting choice for comparison with the USA on account of the relatively greater cultural similarity between the USA and the UK compared to the rest of the EU. However, as we shall see, much if not most of the differences between general European approaches to GM crops and food and US approaches are reflected in the British attitude. Of course the British share the EU's regulatory regime (to be studied in Chapter 5), but there are various areas of relative independence of action, and we can see the ways in which these areas compare with those of the USA.

3 Britain turns sour on GM food

One of the more surprising outcomes of the GM food and crops story is the extent to which the British have rejected, or at least been sceptical of, GM food and crops. It surprises some radical green Euro activists (Leskien interview – 2002) and it certainly depresses US interests who see the UK as the friendliest face in Europe. The UK government has consistently tried to accommodate agricultural biotechnology interests, and Prime Minister Blair has consistently spoken out in favour of the technology. However, the government found itself unable to prevent a moratorium at a European level. It also found it necessary to negotiate a pause in the commercialisation of GM crops in the UK while trials into the relative impact on wildlife of herbicide-tolerant GM crops and conventional crops were conducted. These trials delivered bad news for the GM crop industry. The government's own advisory committees have adopted a precautionary approach. Finally, and most importantly, the British government has no control over the fact that GM food has been driven off the supermarket shelves.

I shall analyse how these and other outcomes have transpired by deploying the theoretical models that have been discussed in Chapter 2. Broadly speaking this involves performing three tasks. First I shall look into a factor which is widely held to be an important background influence to the GM food and crop debate, that of the general 'crisis' or set of crises surrounding food policy. Second, I shall deploy the 'discourse-paradigm' approach which consists of the following type of study. If we want to understand the nature of GM crop and food decision making we need to study the nature of key discourses and the nature of the scientific advisors concerned with the GM food and crop decision making structure. The regulatory discourses determine what is and what is not true (for the purposes of regulatory decisions). These include interpretations of uncertainty on different issues and what experiments, tests and measurements are necessary (or are not necessary) to divine the knowledge that serves the dominant truths. These discourses are expressed in both scientific and non-scientific language and are based on common cultural attitudes, such as the attitude to food. We also need to investigate how changes in discourses have affected decisions, and we need to associate these discourses with changing practices including the changing membership of the scientific advisory apparatus.

This leads onto the third task, analysis of how political outcomes occurred. This involves comparing the various discursive regulatory and scientific structures brought to light, and also utilises tools such as network analysis both at the level of governmental policy and at the level of particular alliances of interest groups. I shall analyse the use of storylines, as described in Chapter 2, in order to investigate how the public was mobilised and how the projects of otherwise disparate interests were co-ordinated. I want also to have a look at the extent to which government efforts to promote wider debate live up to the objectives implicit in a deliberative approach to democracy, something I shall do in Chapter 6.

In doing all of this I want to discuss what rules may be significant as ways of comparing the institutions governing regulation of GM food and crops in the UK, EU and the USA. This follows on from the discourse analytical theoretical framework discussed in Chapter 2. For example, analysis of the discourses which underpin UK regulatory scientific committees, rules about the membership of such committees and the nature of policy networks which determine policy on GM food and crops can be compared with broadly similar analyses applied to the USA and EU case studies in later chapters.

My account begins with an analysis of the BSE and food crises in the UK. If we are going to understand the belief systems which surround both the advisory scientific machinery and also the institutions involved in the politics of GM food and crops we need to have a look at the 'food crisis'. My discourse analytical approach also involves path dependence and the responses to the GM food and crops issue that have been influenced by the BSE saga. Hence I shall begin with a discussion of the BSE crisis and other food crises.

The BSE and food crisis

BSE (Bovine Spongiform Encephalopathy) was first diagnosed in cattle in 1985. This created concern that the disease could spread to humans, and this was a factor in the setting up of a Commission under the chairmanship of Richard Southwood. The Southwood Report, published in 1988, said that the risk of transmission to humans was 'remote' although the report did recognise that it would be a decade before 'complete reassurances could be given' (Southwood 1989 cited by Reilly and Miller 1997: 240). The spokespersons for the Ministry for Agriculture, Fisheries and Food (MAFF) emphasised the conclusion that there was little to worry about, but did not publicise Southwood's more cautionary warnings (Reilly and Miller 1997: 240).

The government's position was attacked by dissident scientists, such as

Richard Lacey, who argued that there was a very good chance of transmission to humans. Support for his views increased when, beginning in 1990, cases of a BSE-like disease emerged in cats. Indeed the very basis of the government's assessments seemed to shift in the sense that 'By 1994 the Spongiform Encephalopathy Advisory Committee (SEAC) evaluated the risk of transmissibility to humans as remote only because precautionary measures had been put in place' (Phillips 2000a). When in March 1996, the government admitted that the disease had spread to humans there was an almighty panic and for some time British beef became virtually unsaleable in British shops. A European ban soon followed. Although British opinion railed against EU states, particularly France, who were reluctant to lift the ban, this may be understandable given the fact that for some weeks even British consumers did not want to eat British beef.

The Phillips Inquiry (set up by the government) into the BSE debacle commented that:

> When on 20th March 1996 the Government announced that BSE had probably been transmitted to humans, the public felt that they had been betrayed. Confidence in government pronouncements about risk was a further casualty of BSE (Phillips 2000b) . . . Throughout the BSE story, the approach to communication of risk was shaped by a consuming fear of provoking an irrational public scare. This applied not merely to the Government, but to advisory committees, to those responsible for the safety of medicines, to Chief Medical Officers and to the Meat Livestock Commission.
>
> (Phillips 2000c)

The Phillips Inquiry stopped short of actually saying that the government had given priority to protecting commercial farming interests over and above dealing with food health concerns. However, one academic analyst, having discussed the failure of the government to take early action to order all cattle with BSE to be destroyed and ban the use of offal commented (writing in 1991): 'This apparently demonstrates MAFF's continued willingness to side with the producers if there is doubt' (Smith 1991: 252).

It does not need a vast jump of imagination to understand how British consumers come to be so sceptical about the assurances on GM food given by the government and its scientific advisors, given this history and the evident discourse of mistrust that seems to exist among the public towards official pronouncements on food health issues. Indeed, as will be discussed later, research conducted on British public attitudes confirms that BSE appeared, at least around 1999, to be the leading reason given for scepticism over GM food by members of the public (Petts *et al.* 2001: 35). But is BSE simply the cause of scepticism over GM food, or are there other factors involved? This is a good example of the problem of looking for clear causation from behaviouralist studies. Because an opinion survey says that people will not eat GM food because of the BSE crisis, does that prove BSE is the cause of the refusal to buy or license GM food and crops? Or are people merely responding with the latest piece of evidence to justify a largely pre-existing stance? It is very difficult to answer such questions, but we can certainly look into the context to see just how clearly this 'cause and effect' link is undermined.

It may be argued that it is as much the underlying attitude of mistrust, evident in the 'risk' society described by Beck, that makes people more concerned about the problem in the first place. This is not to say that such fears are misplaced, merely that what is and what is not likely to cause public concern has shifted over time. Smoke and sulphur pollution were much higher in London in the nineteenth century compared to 1950, but it was only in 1950 that 'pea-souper' events provoked widespread demands for political action leading to the 1956 Clean Air Act.

The statements quoted above from the Phillips Report are very much in keeping with the theoretical scripts written by authors such as Beck and Irwin (discussed in the previous chapter) about changing attitudes to risk and their growing mistrust of official experts. These theoretical accounts were written before the denouement of the BSE crisis was reached in March 1996. Moreover, the research on trust in information sources indicating lack of public trust in governmental and official scientists quoted in the last chapter was conducted before 1996. Of course it is quite possible to argue that the BSE crisis was merely the latest in a series of food crises and declines in confidence in regulatory authorities in the EU compared to the relative (food) regulatory tranquility in the US, as is argued by some.

The notion of food crisis or food scare was well in use in the 1980s on issues such as salmonella in eggs. The salmonella-in-eggs crisis erupted with particular vigour when a government minister, Edwina Currie, warned against eating raw eggs because most eggs were infected with salmonella. Consumers reacted with panic and egg sales plummeted, for a while. In fact Currie was forced to resign after incurring the chagrin of the National Farmers Union (NFU). Listeria also gained a high profile in the UK media during the 1980s (Smith 1991: 252). Although, as will be mentioned later, there have been much the same problems with salmonella and listeria in the USA, there was not the same level of public concern.

It is, however, undeniable that the flow of the first (US) imports of GM food into the EU (towards the end of 1996) occurred against the backdrop of the BSE crisis. Then again, the regulatory regime which had already been established in the EU to deal with GM crops was already rather different from that of the USA. I shall discuss such differences later in this chapter and also in succeeding chapters on the EU and the USA. This difference between UK/EU regulation of GM crops is an issue in the analysis, which I set out below, of the nature and changes in the scientific regulation of GM crops in the UK. I use the 'discourse-paradigm' approach that I outlined in Chapter 2.

Discourse-paradigms in UK GM crop/food policy

I want to look at the belief systems which underpinned the 'science' of GM regulation in the UK. In doing so I need to emphasise that the legislative framework, also the main parts of the process of authorising testing and commercialisation of particular GM food and crop products, is done at an EU level. This includes policy on the labelling of GM food. The UK government does, nevertheless, make recommendations to the EU concerning GM products. It can also sponsor voluntary agreements on GM crops and food. I shall focus on these areas of domestic governance in this chapter.

I want to focus mainly on the Advisory Committee on Releases into the Environment (ACRE). This was specifically established under the terms of the UK's 1990 Environmental Protection Act (and paralleled EU legislation on GMO releases) to advise the government on the deliberate release of GMOs into the environment. ACRE also deals with potential releases of non-indigenous plant and animal species, but these form only a small portion of ACRE's business. Yet, as we shall see later this parallel with non-indigenous species is highly significant.

There are various other scientific advisory bodies which advise the UK government on aspects of GM food and crops. In particular there is the Advisory Committee on Novel Foods and Processes, but this deals with many issues, such as proposals for food with iodine supplements or cholesterol reducing ingredients and only a minority (albeit a substantial minority) of the time with GM food. An interesting fact to note (and something which will be discussed at greater length in Chapter 4 on the USA) is that in the USA there is no similar body and that GM food is not considered to be fundamentally different to food from conventional crops. Like the Advisory Committee on Novel Foods, the Committee on Animal Feedingstuffs spends the bulk of its time dealing with non-GM feed issues. The Advisory Committee on Pesticides (ACP) has also dealt with GMO issues in that the ACP has to give clearance to any pesticide uses that go along with GM crops, but again, this constitutes only a minority of the ACP's business.

The scientific advisory bodies effectively decide the stances of UK representatives on regulatory bodies within the EU and other international organisations. Its most important legal focus are the terms of EU legislation on GM crops contained in the Directives 90/220/EEC, later revised by 01/18/EC, the nature of which will be discussed in Chapter 5. ACRE and the other scientific advisory bodies operate under the cloak of science, but as we can see, their 'science' is limited by the belief systems, the discourses, underpinning the scientific communities which comprise the committees dealing with GM food and crop regulation.

The advice given to officials and ministers about GM food and crops will either be about a product that is actually being tested in the UK or about dossiers supplied by another EU country where the tests are being carried out. The host state for the tests is known as the 'raporteur'. If the UK is the raporteur the Advisory Committee on Releases into the Environment will present an applicant (usually a biotechnology company) with a series of questions to answer. These questions will often require experimentation to generate the required information which is then compiled into a 'dossier'. If the 'information' so generated satisfies the advisory committee, then authorisation will be recommended, otherwise the dossier will be rejected and more information will be requested. At this point the applicant may withdraw in the knowledge that it cannot provide the required data. This process highlights an important aspect of the practical application of the discourse-paradigm theory. The questions, the requests for information, posed by the scientific advisory committee form the link between its underlying discourses on risk assessment and the positivistic experiments that are required as means of validating knowledge under those discourses.

My approach, in discussing the discourses underpinning this system of scientific advice, is to pick out particular areas of belief system that are important in the comparison between the USA, EU and the UK. Some of these discourses have changed as a result of the 1996–1999 'crisis' in the regulation of GM food and crops. I shall begin with a description and analysis of these discourses as they existed in 1996, just prior to the controversy and changes in UK GM crop policy. I shall then discuss the key changes in these belief systems. I shall demonstrate (later) that these belief systems, which act to guide the selection of, and design of, positivistic experiments, are closely related to changing policy/political/governmental management policies. It needs to be emphasised that the scientific regulatory bodies operate very closely within the remit of EU law. The bulk of the deliberations of these bodies (certainly including ACRE) are concerned either with acting as raporteurs for proposals for GMO release licenses or with commenting on proposals made by other countries. I shall say much more about the EU policies on GMOs in Chapter 5.

Perhaps the most fundamental part of the belief system underpinning ACRE is that it exists as a distinct body at all. Implicit in its existence is the belief that GM crops are distinct from conventional crops and that they present a fundamentally different set of problems compared to new plant varieties developed by traditional means. Although scientific advisors in the USA insist that their assessment of GM food and crops is at least as rigorous as in the EU, the European notion that GMOs are distinct lends a somewhat different colour to arguments, for example, about whether GM food should be labelled. As will be discussed in Chapter 5, the distinctive way that GM crops (and also GM foods) are regulated in the EU compared to the USA is one of the key institutional contrasts between EU and

US GM food and crop regulation that pre-date the height of the BSE 'crisis' which occurred in the spring of 1996.

It is profoundly significant to this comparative study that GM crops have, since the establishment of ACRE in 1990, been regarded as being fundamentally different to conventional crops. GM crops are regarded as being analogous to 'exotic' foreign species, and hence releases of the exotics are also regulated by ACRE. This marks out a contrast with the US regulatory system (Jasanoff 1995). More will be said of this in Chapters 4 and 5 on the USA and the EU but it is relevant here to quote a passage from the 1989 Report by the Royal Commission on Environmental Pollution (RCEP) which preceded the setting up of ACRE. The report said that:

> If a genetically modified organism were released into an environment in which the unmodified organism was not native, experience with exotics could be highly relevant. Even if the release were into the native environment, the unmodified organism might well be adapted through natural selection to survive in that environment and the genetic manipulation might, possibly deliberately, upset the ecological balance that normally helped to limit the population growth of the unmodified organism.
>
> (Royal Commission on Environmental Pollution 1989: 21)

The RCEP report discussed several examples where exotic species had upset ecological balances. Examples included rabbits introduced into Australia, invasive cross-breeds of rhododendron bushes in the UK and the parasites introduced into Hawaiian farming systems which were supposed to control crop-eating moths but which also destroyed butterfly caterpillars and, indirectly, the wasps and birds which used caterpillars as a food supply. The comprehensive (and, relative to conventional crops, distinctive) GM crop regulatory system set up by the EU in 1990 (which is discussed at length in Chapter 5) was a measure that was demanded by member states, including the UK.

A second key part of the belief system underpinning the GMO regulatory system was the attitude to what the wildlife conservation groups call 'biodiversity' issues. There was, until 1998, an apparent chasm in the eyes of wildlife protection groups, between the Advisory Committee on Pesticides (ACP) and the Advisory Committee on Releases into the Environment (ACRE). The ACP would only consider the toxicity aspects of any pesticide that was used with GMOs while ACRE would consider only the gene transfer issues. Ecologists complained about the 'reductionist' nature of this approach which ignored the effects of the herbicide regime system as a whole. The ecologists insisted that their preferred 'holistic' attitude should look at the impact of a particular GM/herbicide regime, as a whole, on wildlife. What changes in biodiversity did the system produce? If, as the biotechnology companies claimed, the herbicide regime increased the degree to which weeds in the crop fields were eliminated, would there not be fewer weeds and thus less food for birds and other wildlife? This was the question to which groups such as the RSPB, and English Nature (the government's own nature watchdog) wanted answers. The experiments commissioned as a result of the requests for information issued by ACRE and its sister committees in other EU states would not produce answers to this question.

Before the autumn of 1998 the dominant discourse on ACRE interpreted environmental risk to involve 'safety' dangers of GMOs to the direct health of people and other animals/plants. It also involved the possibility of economic harm. Wider issues were discounted. As John Beringer, the former Chairman of ACRE commented in the foreword to the 1996/1997 Report:

> The responsibility of advisory committees, such as ACRE, is to develop scientific procedures for assessing risks, consider risk assessments and advise whether the GMOs are at least as safe as the parents from which they are derived. Social, ethical and other issues arising from the technology should be debated elsewhere by those with the appropriate competence ... ACRE's remit is restricted to the consideration of the safety of GM crops per se in the environment, and does not extend to consider the wider issues such as the environmental risks or herbicide use on GM crops.
>
> (Beringer 1998: 1–4)

It seemed to ecologists to be rather ironic that potential threats to biodiversity were specifically ruled out of consideration by the body set up to advise on the impact of GM crops on the environment. Since ACRE is the official advisor to the government on authorisation for GMO releases the effect of the dominant policy discourse was to reject not only wider social and ethical issues from consideration but also, implicitly, the research demands made by what were then 'dissident' wildlife scientists. These demands involved research into the impacts of GMOs on food availability for birds and other ecological concerns. The lone green representative on ACRE, Julie Hill, articulated environmentalist concerns, but the 1996 ACRE Annual Report specifically discounted her main concerns (Beringer 1998: Chapter 3). As we shall see later, this discourse changed when the government altered its stance in order to deal with mounting controversy over GM food and crop policy.

A third important issue that is relevant here are the discourses on gene flow, or genetic pollution as the radical environmentalists put it. There was extensive discussion (albeit based on limited experimentation) of the gene flow issue by ACRE in the pre-1998 period. Environmentalists argued that gene flow to related species could have unforeseen ecological consequences. In the case of herbicide tolerance, there would be an increase in levels of tolerance for common herbicides which could lead to an increase in use of other more toxic herbicides to deal with plants that had acquired resistance through gene flow. ACRE considered such arguments, but rejected them saying:

> The development of tolerance in target weeds or volunteers will result in the loss of efficacy of a product in a particular situation but would not be a risk to human health and safety or the environment. If there was some loss of efficacy of a herbicide as a result of a small propor- tion of target weeds developing tolerance, the application of the her- bicide would not result in harm to the environment occurring because: the application would still effectively control most other target weeds; and, the application *per se* would not cause harm to the environment.
>
> (Beringer 1998: 10)

MAFF, whose views seemed to be predominating in ACRE, regarded gene flow as a farm management issue, and not a matter of ecological import- ance (Levidow 2001: 854; Curtoys 2001 – interview). ACRE recognised that 'The definition of "harm to the environment" which is used in the Environmental Protection Act 1990 includes harm to man's property' (Beringer 1998: 5). However, ACRE, along with MAFF, asserted that any harm resulting from cross pollination could be avoided by voluntary agreement on management practices rather than regulation. Following a consultation exercise in 1997, MAFF laid the ground for the setting up, in the following year, of an organisation responsible for ensuring the proper management of GM crops. This body, which consisted of various biotech- nology and agricultural interest groups, had the acronym SCIMAC and I shall discuss it later in this chapter in the section on networks. English Nature and the RSPB were inclined to pursue the issue of gene flow more vigorously.

Of course a further construction of the 'gene flow' or 'genetic pollu- tion' issue was as a food health or consumer one. If people were con- cerned that their food was going to be 'contaminated' by GM food, then gene flow to neighbouring conventional or organic crops could be of major concern. Organic farmers and consumers of organic food were especially sensitive to such issues. Yet consideration of such issues was ruled out by the ACRE discourse which saw them as food health issues properly dealt with by the Advisory Committee on Novel Foods and Processes. However, there was a possible inconsistency here, because the ACRE discourse still recognised 'harm to man's property' as an environ- mental problem, discussed in the context of whether the spread of herbi- cide resistance to adjacent fields would cause problems for the farmer of the adjacent field (Beringer 1998: 5).

Organic farmers may have played up the 'contamination' issue to promote the case for organic food, but they still clearly had a case to argue that 'contamination' from GM crops was something that their consumers were very keen to avoid. Of course, whether this 'contamination' would be at levels higher than normally associated with pesticide spillover (from neighbouring conventionally grown crops) onto organic crops is a different matter. The Soil Association, which promotes organic farming, commissioned a scientific report on the possibilities for cross pollination (Emberlin *et al.* 1999) and the issue of contamination increasingly became a big issue in the press (Lean 2002). However, the dominant ACRE discourse was to remain (up until the time of writing) essentially unchanged on the issue of gene flow, with the proviso that further research was needed into the implications for wildlife conservation.

A fourth important (and also controversial) part of the belief system underpinning the scientific regulation of GM food and crops concerns antibiotic resistant marker genes. As I discussed in Chapter 1, earlier designs of GM crops used antibiotic resistance as a convenient means of producing GMOs, and the US regulatory authorities did not object. However, in 1996 (and this certainly coincided with the BSE crisis) the Advisory Committee on Novel Foods and Processes, in common with the food science advisors from most other EU states, rejected a commercial application for a type of GM maize that contained the DNA for ampicillin, a commonly used antibiotic. A rare division inside the UK's scientific regulatory apparatus occurred because ACRE was content to support authorisation of the maize.

ACRE decided that the chances of the spread of the antibiotic resistant gene to other bacteria (chiefly in cows' guts that were given antibiotics) were low. Beringer commented:

> We advised that there was a risk of transfer of the antibiotic resistance gene from the GM plant to bacteria (making them resistant to the antibiotic), but that against the very high background of ampicillin/ penicillin resistant bacteria already present in humans the very rare occurrence of transfer of the gene would not be a clinical problem. As is well known, our advice conflicted with that of the Advisory Committee on Novel Foods and Processes and the UK voted in Brussels against the introduction of the maize.
>
> (Beringer 1999: 4)

Here we see two quite different judgements on uncertainty produced by two sister scientific committees in the same country at the same time. If this can happen then it is clear that judgements about scientific uncertainty must surely be regarded as cultural phenomenon resting on differing discourses concerning uncertainty issues rather than positivistic scientific judgements!

The fifth, and final issue of controversy concerns the possibility or unintended consequences of DNA transfers. Some DNA that is inserted into the recombinant gene is not directly associated with the desired attribute, for example, herbicide resistance. Critics of GM food technology have argued, amongst other things, that the imported DNA may be placed in a sequence that may produce unintended consequences, or that extra DNA imported into the organism may have unintended consequences. In the year 2000 the Advisory Committee on Novel Foods and Processes re-opened an assessment of Monsanto's 'roundup ready' (herbicide tolerant) soya beans which had been cleared for import and processing by the Committee in 1994. In those days there had been no consideration of any DNA transferred that was additional to that which conferred herbicide tolerance. However, further 'molecular characterisation' was carried out to determine what extra DNA had been transferred and whether it was functional (that is, having an effect). Analysis did discover some extra DNA but it was found not to be functional (Advisory Committee of Novel Foods and Processes 2001: 13; Food Standards Agency 2000). Now I want to examine the discursive reconstruction of what is regarded as good science of GM food and crop risk assessment.

Changes in discourse-paradigm on GM food and crops

The first change in discourse that I want to discuss is concerned with food safety and DNA transfer. The significance of the case of the re-examination of DNA structure of Monsanto 'roundup ready'[1] beans to this analysis concerns how changes in scientific assessment procedures were triggered by changes in social and political attitudes. The discourse underpinning the UK's scientific regulatory machinery for assessing GM food and crop technology changed, from one of not regarding the issue of additional DNA sequences as being significant (in 1994) to regarding the issue as having significance (by 2000). Of course, the social context of this change was increases in public concern about GM food and crops, and controversies such as the Pusztai affair (see Chapter 1) whereby critics of GM technology claimed that the alterations in the genetic structure of GM food was rather more complicated with more scope for unintended consequences than the biotechnology companies claimed.

More sophisticated molecular characterisation procedures have been developed, and regulatory committees now want to know, in the products that are being assessed, whether there are any extra DNA strands being transferred and whether they are functional. Discursive change means that different questions are posed for risk researchers which in turn demands that new scientific experiments or analyses be conducted. The choice of which positivistic tests to apply is thus dependent on the nature of the discourse. The nature of the dominant discourse underpinning the scientific experiment is itself affected by social and political debate. We can see from this example that it is in practice (as explained in theory in Chapter 2) very difficult to separate out the belief systems that underpin science from the belief systems that are dominant at any one time among the public, or in government departments and regulatory bodies. Saying that a particular scientific technique is value free means little because techniques are only used in a context, and that context will not be value-free. The key issues are: *which* scientific, positivistic techniques and experiments are deployed; whether the data that they produce is or is not considered to be significant; and, most crucial of all, how that data is interpreted.

In fact the most politically significant alteration in the dominant discourse underpinning the UK's GM food and crops risk assessment machinery was not directly connected with food safety, but rather with biodiversity and wildlife protection. This is the second area of discursive change I want to discuss. I mentioned earlier that key concerns of wildlife protection groups about the impact of GM crops on the availability of food for wildlife were not being taken into account by ACRE when considering whether to give commercial approval to particular GM crops. However, this was to change in 1998.

Following considerable political controversy in the 1998 and 1999 period ACRE was reconstituted. The reorganisation took effect in June 1999, although it implemented the settlement with the wildlife protection groups which took shape in the autumn of 1998. A new series of experiments, the Farm Scale Evaluations, often called the FSE Trials, was established. Their function was to test whether various herbicide tolerant GM crops, that were awaiting approval for commercial release, reduced the availability of food for wildlife, especially birds. They were called 'farm scale' because conventional and GM variants of particular crops were planted at the same time so that the impact of the crops could be observed in the same conditions. Four herbicide tolerant crops were subjected to these studies, comprising a type of sugar beet, a variety of maize (for animal consumption only) and two types of oil seed rape.

The ACRE discourse underwent a significant change, as indicated by the following statements by the Chairman of ACRE:

> I am delighted that the Government has recognized that wider issues relating to the release of genetically modified organisms should be considered in the context of agriculture as a whole ... English Nature, the Government's advisors on wildlife conservation, are concerned that the introduction of some types of GM crops into commercial agriculture may prejudice commitments to reverse declines in farmland wildlife. It is important that this issue is explored fully and that research to investigate the impact of GM crop management in agriculture on farmland wildlife is carried out.

(Beringer 1999: 5–15)

It is clear from this that the discursive change was politically promoted, and that scientific regulatory committees are, if this example is at all typical, mere poodles for political expediency. Once again we can see that changes in discourse inspire new questions which are dealt with by conducting new types of experiment and investigation. Positivism, in this sense, is a subset of discourse.

This particular discursive (and thus scientific) change was welcomed by English Nature and the RSPB. However, it did not satisfy a range of other interests, including the radical environmental groups, organic farmers and food/health editors of tabloid newspapers, the latter being keen to speak to their reader's worries about food safety concerns. The radical environmentalists attacked the research, for example, because it involved a comparison with conventional farming rather than organic farming (their preferred strategy), and because it involved open air releases (Holden interview – 2001). The trials could not, the radicals argued, take account of small, cumulative, long term changes or differences between practices under experimental conditions and what farmers do in normal operations (Riley interview – 2001). Scientists working for English Nature responded by saying that organic farming, however desirable, could not hope to replace conventional farming in the medium term and that herbicides used with some GM herbicide tolerant crops (mainly maize) were actually preferable to toxic, long term residue herbicides such as atrazine used in conventional farming (Johnson interview – 2001).

Notions that it is impractical to concentrate solely on a transformation of conventional into organic farming are political judgements. Science may help answer certain questions, but a given discourse demands only that some questions be asked and that other questions are irrelevant, misleading or mistaken. Often, uncertainty is also interpreted according to political expedience. Some would argue that the three-year-long Farm Scale Evaluation GM crop trials cannot say anything about long term impacts. Here scientists have to make a judgement that takes into account commercial issues. As the Chief Scientist for the RSPB commented: 'We could have argued that we should have a ten year research programme before commercial release, i.e. Farm Scale trials to go on for even longer, but the biotechnology companies would simply give up and go away' (Gibbons interview – 2002). The risk/benefit judgement shapes the design of the experiments, and the nature of the risk/benefit judgement is decided through discourse.

The results of the FSE Trials proved to be largely unwelcome for the biotechnology companies involved. Eight papers discussing the results of the FSE Trials were published by the Royal Society in October 2003. The results were delivered too late in the preparation of this book to be discussed at length, but in brief the research suggested that many herbicide tolerant GM crops significantly reduced the availability of food for birds. The quantities of available seeds and insects were generally reduced in the trials on oil seed rape and sugar beet. The numbers of bees and butterflies were also reduced. In the case of the maize trial the outcome was more positive for the GM variety since the GM herbicide regime was being compared with the use of atrazine which has a toxic reputation. However, many observers questioned the value of this comparison since the EU plans to phase-out use of atrazine.

ACRE confirmed the findings of the FSE Trials in January 2004, although their findings were widely reported as giving the green light to the herbicide tolerant maize. Given that this type of maize had already been given authorisation at an EU level this seemed to herald the commercialisation of this GM maize. However, there is widespread doubt that a sufficient market exists for the fodder maize that would be grown.

The third discursive change which I want to highlight concerns attitudes to GM crops using antibiotic resistant marker genes. I mentioned earlier how ACRE, in a notable disagreement with the Advisory Committee on Novel Foods and Processes, had recommended to the EU that a variety of GM maize be given a commercial license for import and processing. However, following the reconstitution of ACRE the discourse saying that antibiotic resistant marker genes posed no significant threat was overturned. In the year 2000 ACRE rejected a dossier supporting an application for the importation and processing of genetically modified flax. Despite the fact that the flax was to be principally used for industrial purposes, ACRE commented that:

> Some of the pressed cake, after oil extraction, could, however, find its way into animal feed ... Unnecessary (DNA) sequences (in the flax genome) include a block of three antibiotic resistant genes ... We concluded that there was a small but real risk of transfer of the antibiotic resistance genes to bacteria, particularly in the gastrointestinal tract of animals.
>
> (Gray 2001: 19–20)

In the case of an application for commercialisation of herbicide tolerant fodder beet from Monsanto, ACRE only agreed to recommend acceptance after they were content 'that a functional streptomycin antibiotic resistance gene was absent' (Gray 2001: 19). We can see here how the dominant discourse on uncertainties associated with the use of antibiotic resistant marker genes had changed from regarding the 'rare' risk of gene transfer in animal guts as not being a major cause for concern to regarding the 'small but real' chance of gene transfer a sufficient cause for concern to merit the rejection of the dossier.

A fourth area of discursive change concerned the issue of genetic pollution. In fact the dominant discourse that this was a farm management issue rather than an environmental problem remains largely in place. However it has been eroded to the extent that post-cultivation monitoring of problems such as herbicide tolerance are part of ACRE's recommendations and requirements for research.

In negotiations held with the government in October 1998, English Nature and the RSPB successfully demanded that some studies on gene flow would be included in the Farm Scale Evaluation Trials which were being set up to test the biodiversity impact of some GM crops (Gibbons interview – 2002). Later on, English Nature published a report which discussed the dangers of 'gene stacking' whereby if a number of GM varieties of, say oil seed rape, which are resistant to different herbicides, are grown, then multiple resistance could develop (Orson 2002). This has developed to a certain extent in Canada (see discussion in Chapter 1). The phenomenon of gene stacking was associated, popularly, with so-called 'superweeds'. English Nature linked gene flow to the biodiversity issue, saying that if there was 'gene stacking' then farmers might respond by using more (perhaps virulent) herbicides to control the 'superweeds', a practice which might reduce food for wildlife. Although the gene flow issue has not resulted in any cases of ACRE refusing to recommend GM crop authorisations, it has influenced the conditions recommended by ACRE for such authorisation. In particular, since 1998 permits for trials explicitly mentioned the need to do monitoring for two years to study the emergence of 'volunteers', that is plants which re-appear in following years with the acquired trait. In addition a study on the patterns of the spread of herbicide resistance to crop margins of fields was included in the FSE Trials.

So far I have discussed the ideas underpinning ACRE in terms of Foucauldian discourse. However, as I have noted in Chapter 2, we need to consider the role of agents, something which Kuhn took into account when he discussed how a 'community' of scientists formed the basis for acceptance of a 'paradigm'. So I want to have a look at how the community of scientists on ACRE changed when the discourse which underpinned ACRE altered at the end of 1998.

The changing ACRE community

As the controversy over GM food and crops became more intense towards the end of 1998 and the beginning of 1999 there was criticism that the membership of ACRE was skewed towards biotechnological interests. Even the tabloid press joined the attack on ACRE with the Daily Mail claiming that eight of its members 'are involved in the biotech industry' (Hinscliff 1999). Friends of the Earth claimed that ten out of 16 members of ACRE 'are either directly employed by, or receive funding for research or other work, from the companies which want to market genetically modified crops' (Brown 1999).

ACRE's membership was reorganised on the basis that members are recruited because they possess certain required disciplinary skills rather than because they represented particular interests. At the same time as this change a rule was introduced which barred ACRE members from having commercial contracts with biotechnology companies, although this rule change did not apply to the Advisory Committee on Novel Foods and Processes which was responsible to MAFF rather than the DETR.

The changes in the disciplinary balance of ACRE clearly parallel the discursive changes that I have discussed earlier. The previous Chairman, Professor John Beringer (a microbiologist) was replaced by Professor Alan Gray (a plant ecologist), and there was a general shift among committee members away from domination by biotechnology/genetics, food health and conventional farming expertise towards expertise in plant ecology, insects, sustainable farming and biodiversity. In addition an ACRE sub-group on wider biodiversity issues was set up, and this sub-group was dominated by wildlife and conservation interests. We can see from the contrast between Tables 1 and 2 (below) that before the reorganisation there was a majority of ACRE members associated with biotechnology and conventional farming interests, but following reorganisation there was a majority of specialists dealing with ecological and biodiversity issues.

In Tables 3.1 and 3.2 below I have contrasted the numbers of different types of expertise according to four categories. Because of the reduced numbers on the new committee (13 names listed instead of 16) I have put percentages in bold to sharpen the contrast.

The ACRE Sub-Group on wider biodiversity, which consisted of 12 members, included four scientists from English Nature and the RSPB as well as other members with expertise in ecology. In addition, the Scientific Steering Committee of the Farm Scale Evaluation Trials which, as mentioned earlier, were intended to test the impact of herbicide resistant GM crops on wildlife, consisted mainly of ecologists from organisations such as the RSPB and the Game Conservancy Trust.

What is of some interest is that despite the fact, as we shall see, that the public scepticism over GM crops seems to have been motivated more by food safety rather than ecological concerns, the involvement of food safety expertise on ACRE was actually reduced. That having been said, the precautionary discourse on antibiotic resistant marker genes became

Table 3.1 Types of expertise on ACRE immediately before 1999 reorganisation

Types of expertise	Number of members	Percent of membership
Conventional farming	3	**18.5**
Ecology, plants and insects	4	**25**
Genetics and micro/molecular biology	6	**38**
Food safety and human health	3	**18.5**
Total	16	**100**

Source: ACRE Annual Report No. 4 1996/1997 Annex B pp. 1–3 (Department of Environment, Transport and the Regions, 1998).

Table 3.2 Types of expertise on ACRE immediately following 1999 reorganisation

Types of expertise	Number of members	Percent of membership
Conventional farming	1	8
Ecology, plants and insects	7	53
Genetics and micro/molecular biology	4	31
Food safety and human health	1	8
Total	13	100

Source: ACRE Annual Report No. 7 Annex C pp. 61–63 (Department of Environment, Food and Rural Affairs).

dominant on ACRE, so we can assume that the now dominant biodiversity discourse was also associated with a precautionary discourse on antibiotic resistant marker genes. Moreover, the Advisory Committee on Novel Foods and Processes seemed perfectly able to deal with such issues. However, the dominant food safety discourses on both committees did not embrace the more drastic 'food threat' discourse with which the radical environmentalists associated themselves. An understanding of how the biodiversity interests gained prominence in ACRE rather than food health or radical anti-GM discourses is discussed later.

It may be argued that the quite obvious political orchestration of the membership and terms of reference of ACRE undermines use of Kuhn's paradigm concept, which, as I pointed out in Chapter 2 implies that paradigms cannot be chosen by scientific communities which are selected by government. Yet a careful examination of independent scientific reports of the Royal Society, which acts as the speaker for the views of the UK's scientific community (or at least its establishment) reveals quite clear support for the UK and EU approach to regulation of GM food and crops compared to that of the USA. The Royal Society has not produced a report covering biodiversity issues, but it has produced reports on food safety issues. The Royal Society supports EU and UK policies favouring the labelling of GM food, opposes the presence of antibiotic market resistant genes in GM food and advocates thorough safety assessments of GM food. Indeed a recent report went so far as to advocate that 'We suggest that the Food Standards Agency considers whether the post-marketing surveillance (of GM food) should be part of the overall safety strategy for allergies, especially of high risk groups such as infants and individuals in "atopic" families' (Royal Society 2002: 11). The Royal Society also acted as the publisher of the papers on the results of the FSE Trials.

Hence there is strong evidence to suggest that the discourse-paradigm that I have described does in fact broadly reflect the views of the UK's scientific peer group. The government has been induced by pressure from both public and scientists, who were immersed in public unease about agricultural technology, to alter that structure. Perhaps though, it is not so much a matter of the scientists influencing society as the influence of science by society. I shall move on to demonstrate how changes in the policy networks underpinning governmental decision making on GM crops corresponded to changes in the discourse-paradigm outlined above.

Policy networks and GM crops

At the time when the first imports of GM food were arriving in the UK at the end of 1996 the GM crop policy machinery was mostly directed by the Ministry of Agriculture, Fisheries and Food (MAFF). It was busy preparing the ground for what was anticipated to be the imminent commercial cultivation of GM crops in the UK. This policy machinery constructed issues such as gene transfer of herbicide tolerance traits to conventional crop plants or wild species as economic issues to be tackled by better farm management. This was reflected in the discourse underpinning ACRE that I have discussed earlier.

After much discussion a body, with the unromantic acronym of 'SCIMAC', was established to set guidelines for the commercial cultivation of GM crops. SCIMAC stood for 'Supply Chain Initiative for the Management of Modified Agricultural Crops'. It consisted of representatives of the main industrial interest groups concerned with GM crop commercialisation, that is the biotechnology companies, crop protection companies, the National Farmers Union, seed producers and farm produce distributors. These interests had dominated decision making on GM crops before the setting up of SCIMAC. Together with MAFF these commercial interests formed what can be described as a 'policy community' (Marsh and Rhodes 1992). This means that governmental policy decision making was restricted to a few interests who had a high level of agreement on basic policy principles. Other interest groups with concerns about the issue, but who disagreed with the policy community's underlying principles, were excluded from the policy community. The members of the policy community had close links and depended on each other through mutual resource exchanges. The Department of Environment, Transport and Regions (DETR) was also in the policy community as the sponsoring department of ACRE (which we have extensively discussed earlier). However, until the summer of 1998 the DETR played a largely passive role, and as was discussed earlier ACRE's terms of reference excluded many of the concerns of critics of GM crops.

Sceptics of GM crops included not only radical environmental groups who were effectively opposed to all open air cultivation of GM crops (including open air trials), but also more 'moderate' wildlife protection groups who accepted GM crops provided it could be shown that they would not damage biodiversity. These wildlife groups included the million member strong Royal Society for the Protection of Birds in addition to English Nature, the government's statutory wildlife protection body.

Initially, radical environmental groups and more insider groups such as the RSPB campaigned together against MAFF's approach and demanded a 'moratorium' on commercial cultivations. English Nature sided with this campaign, at first by pressing its case in private. Working in collaboration with other wildlife conservation groups they demanded a set of trials to examine the impact of GM crops on food for wildlife. However, their demands were rejected. External to this debate radical environmental groups organised consumer campaigns to boycott GM foods and they scored a notable consumer success in the spring of 1998 when Iceland, a major supermarket chain, announced that it would exclude GM food products from its own brand names.

In June 1998 English Nature launched a public campaign for its demands, leading the call for a moratorium on commercial cultivation of GM crops. It was in the summer that the DETR began to take the lead on the GM crop issue. The Minister for Environmental Protection, Michael Meacher, is credited with orchestrating the key changes which brought biodiversity considerations to the forefront of GM crop policy (Toke and Marsh 2003). Over the course of the next 14 months all of English Nature's demands were conceded and the wildlife conservation lobby were placed firmly at the centre of GM crop policy decision making. This involved more members (now a majority) concerned with wildlife conservation on a reconstituted ACRE and also a majority of members on a new Scientific Steering Committee whose role was to oversee the 'Farm Scale Evaluation' Trials of the impact of GM crops on biodiversity.

Hence what I would call the 'GM Industry Policy Community' was transformed into a 'GM Environmental Protection Policy Network'. This network incorporated the demands of the wildlife conservation interests, but the radical environmental groups remained outside the network. The policy community therefore became less exclusive and there was less policy agreement. However the membership of the policy network still agreed on two key principles. First, they all supported the notion of commercial planting of GM crops given a proviso that they did not reduce biodiversity. Second, that there was a moratorium on commercial planting until 2003. The policy network therefore occupied a place somewhere between the policy community and the looser issue networks described by Marsh and Rhodes (1992). The changes in the policy network are set out in Table 3.3 below.

The key linkage I want to make here is how the changes in scientific discourses, and the changes in the membership of the scientific regulatory Committee (ACRE) which I described earlier were associated with changes in the constellation of political and economic interests involved in governmental decision making. These two changes, the political changes at the level of policy network membership and the changes in the discourses and membership of the scientific regulatory committees, cannot be separated in any meaningful manner. We can identify members

Table 3.3 Changes in GM policy networks

Membership of GM industry policy community until October 1998	Membership of GM environmental protection policy network from October 1998	Groups excluded from both policy networks
• Biotechnology companies • Crop Protection Association (CPA) • NFU • Seed producers • Farm produce distributors • The groups above were coordinated by SCIMAC (GM crop lobby) from June 1998 • MAFF • 'Pro-biotechnology' majority of scientists on Advisory Committee on Releases into the Environment (ACRE) • DETR (passive)	• SCIMAC (GM crop lobby coordinating biotechnology companies, NFU, CPA, seed producers, farm produce distributors) • DETR (active) • MAFF • 'Independent' scientists on Advisory Committee on Releases into the Environment (ACRE) • English Nature • RSPB • Game Conservancy Trust	• Friends of the Earth • Greenpeace • Soil Association • Action Aid • Other Anti-GM groups

Source: Toke, D. and Marsh, D. 'Policy Networks and the GM Crops Issue: Assessing the Utility of the Dialectical Model of Policy Networks', *Public Administration*, forthcoming, 2003.

of the policy network, for instance ministers and biotechnology companies, as actors that are distinct from the scientists who sit on regulatory committees, but these actors are themselves party to discourses whose shape and dominance has simultaneously shifted at the level of both the overtly political policy network and also at the level of the overtly scientific committees.

Moreover we can see how the nature of the reconstituted policy network (now more concerned with environmental protection than industrial sponsorship) still excluded key interest group concerns, namely those of the radical environmental groups and the principal representative of the organic farming movement, the Soil Association. These groups rejected GM crops and food almost on principle.

Besides using network analysis to conceptualise governmental decision making, we can also use a different sort of network analysis to study the non-governmental alliances and networks which try and shift the policy agenda in their favour. I want to study these networks and their storylines. We can trace three networks which all have their own common belief systems that are distinct in key respects from the other networks.

Non-governmental networks

The GM crop network

This network comprises those interests which were concerned with promoting the commercialisation of GM crops. This network was formally constituted under the umbrella of the 'Supply Chain Initiative for Modified Agricultural Crops' by agreement with MAFF following consultations with government about how GM crops could be commercialised. The aim of SCIMAC, in formal terms, was to draft guidelines for the responsible management of GM crops, although in practice SCIMAC acted as a focus for political lobbying for GM crop commercialisation in general. The groups involved in SCIMAC, in effect, formed the GM Industry Policy Community with MAFF which decided GM crop policy until the reorientation of GM crop policy which occurred from autumn 1998 onwards.

The GM Industry Network was spearheaded by the biotechnology companies. Three companies, Bayer Crop Science (formerly Aventis Crop-Science), Monsanto and Syngenta are involved in the Farm Scale Evaluation Trials. These companies are leading members of both the Plant Breeders Association and the Crop Protection Association, which are two of the five constituent members of SCIMAC. The companies that develop GM crops are also companies whose business are heavily reliant on selling pesticides, (crop protection as the industry puts it), such as herbicides, insecticides and fungicides. Monsanto, Syngenta and Bayer Crop Science are all multinational concerns. The companies' GM business is principally US based, but the biotechnology companies are determined to carry on pressing for commercialisation in Europe because of its influence on the world market. The UK is seen as one of the more potentially fertile areas, in Europe, for the GM crop industry. The Managing Director of Syngenta bases his storyline advocating GM crops by citing Harold Wilson's 1960s call to 'harness the white heat of the technological revolution' (Smith interview – 2001). Roger Turner from the Plant Breeders Association and a SCIMAC Chairman has said that 'Genetic modification is a fundamentally sound technology, which will enable us to produce cheaper, better quality food in more sustainable ways' (Meikle 1999).

The other 'half' of the GM crop network is on the agricultural side. The National Farmers Union, who are anxious to help their farmers acquire the newest, cost effective technology in a way that assures good management, has continued to support commercialisation. However the NFU is also keen to stress its respect for consumer concerns. There is still a market for GM animal feed (currently imported), but with the EU's latest labelling regulations requiring GM animal feed to be labelled, supermarkets may start sourcing their meat products from animals fed on non-GM feed, hence closing this potential market. The fourth member of SCIMAC is the United Kingdom Agricultural Supply Trade Association. Its members include suppliers of produce from arable farming, animal feed, seeds, fertilisers, other chemicals and the associated merchants and road hauliers. The interests of the agricultural suppliers, in addition to the fifth member, the British Sugar Beet Producers Association, are closely allied to that of the National Farmers Union.

SCIMAC has accommodated the government's decision to test the impact of herbicide tolerant crops on the food supply for wildlife, although some biotechnology interests were reluctant to agree to a three-year freeze on commercial planting. Hence the government could not announce a freeze until November 1999. However, the fact that the EU had declared a de facto moratorium on authorisations of GM crop commercial licences, and also the fact that EU legislation on testing the environmental impact of GM crops was being strengthened, persuaded most SCIMAC members that there was little point in resisting this demand for a moratorium on commercial planting. It can also be said that despite the UK government's opposition to the EU moratorium on GM crops it appears to be the existence of this EU moratorium that persuaded SCIMAC to agree to a voluntary pause in commercial planting, so allowing the government a breathing space from the quite heavy public criticism that it was experiencing. The UK based biotechnology companies have reconciled themselves to labelling of GM foodstuffs, but they call for consumers to be given 'choice' by making GM food products available on supermarket shelves (Pearsall interview – 2001).

Following the results of the FSE trials the GM crop network appeared to have lost a lot of confidence in its own future. Both Bayer and Monsanto announced that they were closing their operations in the UK.

The wildlife conservation network

There is a very long list of wildlife conservation organisations in the UK, but the principal ones which have been active on the GM crops issue, and which have actively networked together in pursuit of a biodiversity preservation agenda are English Nature, its sister statutory organisations covering Wales and Scotland, the Game Conservancy Council, the British Trust for Ornithology (BTO) and the Royal Society for the Preservation of Birds (RSPB). Brian Johnson of English Nature advances a 'biodiversity' storyline to advance the cause promoted by the wildlife conservation network:

> Our argument with the biotechnology industry is not that they are the biotechnology industry, but rather with the way that Research and Development is going. They see transgenic technology in agriculture as intensifying agricultural production. They talk about more yield, cleaner farming, more convenient farming which we know are very damaging to biodiversity in Europe. They may be appropriate and sustainable in the US, although I have my doubts, but they are damaging

in Europe. We want to see agricultural biotechnology used in a way that makes agricultural systems more sustainable.

(Johnson interview – 2001)

The central concern of the wildlife conservation network is birds. This network has been particularly alarmed by research showing that common species of birds in the UK have declined by drastic amounts over recent years. For example the number of skylarks in the UK has been cut by half since the end of the 1960s. This development is blamed on the intensivisation of agriculture which has increasingly sterilised fields (Johnson interview – 2001; Gibbons interview – 2002). Other types of wildlife, for example bees, have also been observed to have suffered broadly similar declines over this period (Williams interview – 2002).

This wildlife conservation network is reformist in its attitude to conventional farming, and while it is sympathetic to proposals for support for organic farming, it does not see it as a realistic substitute for conventional farming in the short term (Johnson interview – 2001; Gibbons interview – 2002). The generally pragmatic approach (to GM crop technology) lends itself to collaboration with government and the possibility for accommodation which was achieved in the 1998 to 1999 period. Nevertheless this only occurred after public campaigning by both English Nature and the RSPB. The wildlife conservation network effectively controls the scientific appraisal of GM crops because of the representation of the RSPB, English Nature and the Game Conservancy Trust on the Scientific Steering Committee of the Field Scale Evaluations. The network also dominates the biodiversity sub-committee of ACRE and the majority of members of the main ACRE body are now concerned with impacts on wildlife.

If we deploy the notion of path dependence it is hardly surprising that the government's response to the campaigns against GM food and crops manifested itself through the adoption of the demands of the wildlife conservation network. Wildlife conservation and biodiversity protection is enshrined in the heart of government through the agency of English Nature. The government is committed though its agreement to the 1992 Biodiversity Convention (via the Rio Earth Summit) to protect and enhance biodiversity. Of course 'biodiversity' requires interpretation; the selection of what species and ecosystems need protecting according to a list of priorities is a political and social issue. If government is going to give ground to critics of its GM crop policy it will do so on the terms of those agents who are closest to government, and in this case agents that are actually inside government. English Nature embodies a conservative British environmental tradition which places most emphasis on countryside and wildlife conservation. In doing so it is allied to the million member strong RSPB and other institutions such as the Game Conservancy Trust who have impeccable elite credentials.

The wildlife conservationists have collaborated with radical groups from time to time, especially over the issue of a moratorium on GM crops. The RSPB kept in regular contact with anti-GM crop groups like Friends of the Earth, Greenpeace and the Soil Association, despite the division over attitudes to the Farm Scale Evaluation trials. For example, the RSPB publicly backed demands by the Soil Association for a GM test trial near the Henry Doubleday organic research centre (in the Midlands) to be abandoned in early 2001 (Curtoys interview – 2001).

The radical environmental network

As is the case with what sociologists have identified as 'new social movements' who are opposed to dominant political paradigms, some radical anti-GM groups have a much less formal approach to organisation compared to the political mainstream. This having been said, most radical environmental groups have used the 'Five year Freeze Campaign', formed in February 1999, as an umbrella group. The Campaign advocates a total freeze on both open air testing and import of GM crops and food. It was founded by Friends of the Earth, the Soil Association, Genetics Food Alert, the Soil Association and the Gaia Foundation. Greenpeace decided not to put its name to the Campaign on the grounds that it is opposed to GM crops as a matter of principle. There is considerable exchange of personnel between the various radical anti-GM groups and a considerable degree of agreement on belief systems.

Sue Mayer runs Genewatch, which serves as the research arm of the anti-GM movement, and as Head of Science at Greenpeace (until 1995) she oversaw the first campaigns against GM crop technology in the UK. These began around 1990. She comments that: 'The reason why we were doing it was a combination of the potential risks, the particular way it (GM crop technology) was being applied in terms of intensive agriculture and the dominant political forces behind it to do with ownership and control not being in keeping with green thinking' (Mayer interview – 2001).

The storyline of seeing GM crop technology as a further step of industrialising and centralising control over agriculture builds on existing discourses such as that laid down by Porritt in his influential 1984 text, *Seeing Green*: 'By turning agriculture into agribusiness, by operating according to an urban industrial model of economic efficiency, we have created some staggering problems for ourselves' (Porritt 1984: 101).

Environmentalists have, since the earlier 1960s campaigns against DDT and other pesticides, stressed the importance of low input farming. By the mid 1990s Friends of the Earth and the other radicals had become firm supporters of a strategy of promoting organic agriculture as the main alternative to conventional intensive farming. As I discussed in Chapter 1, there were some voices, at the beginning of the 1990s, who argued that GM crop technology could help organic farming, for instance by developing nitrogen fixing crop varieties, (as an alternative to fertilisers).

However, it was the prospect of external control over farmers through ownership of seed patent right by biotechnology companies that seemed to turn the organic movement firmly against GM crop technology.

There are differences of emphasis among the groups in the anti-GM network. Development groups are opposed to the intensification of agriculture, but focus on the possible loss of farming jobs in Third World countries if GM technology takes hold. Groups like Genetix Snowball and Greenpeace have organised direct action campaigns against trials of GM crops, an activity that is spurned by groups such as Friends of the Earth. The crop pulling exercises, which took off with great regularity in 1998, certainly grabbed the headlines. This was also the year that labelling of GM food took effect in the supermarkets.

However it was the mobilisation of ordinary consumers to call on supermarkets to stop stocking GM food that had the most direct effect in curtailing the commercialisation of GM food in the UK. The groups organised consumers to write in and phone in demands for GM free shopping, and one by one the supermarkets and, importantly, the major food processing companies agreed to source non-GM food supplies. Iceland was the first to announce, in March 1998, that its products would not contain GM ingredients. Later Marks and Spencer's also took a deliberately 'green' marketing stance, but once other supermarkets realised that their 'GM free' competitors would gain competitive advantage, the GM food products rapidly disappeared off their shelves (Kronick interview – 2001; Morris interview – 2001). The battle for the supermarkets was largely settled by the middle of 1999. The only market footholds for GM food products since then have been animal feed and also for derivative products, such as starch or oils. The processing of the food destroys the original DNA. However, the EU's Food and Feed, and Labelling and Traceability regulations come into force in 2004. These require derivative products to be labelled as well as GM animal feed. These changes will make it easier for the supermarkets to source non-GM food products in these areas.

The Consumers Association do not support the radical environmentalist 'anti-industrial agriculture' discourse. Their dominant discourse is 'freedom of choice for the consumer', but their market research makes them feel that the majority of consumers are hostile to GM food and that they want to avoid non-GM food being contaminated by GM food. The Consumers Association has backed demands for labelling of derivatives and urged a moratorium on GM food and crop authorisations until measures have been put in place to counter contamination of non-GM food supplies by GM food products (Consumers Association 2002a). The Consumers Association used strong language to criticise the failure of the Food Standards Agency to support the labelling of derivatives. Peter Jenkins, Senior Public Affairs Officer Consumers' Association, said:

We are bitterly disappointed by the FSA's decision to retain its anti-quated, anti-consumer stance on GM. UK consumers will now have to rely on other European Governments to stand up for their views on GM labelling.

(Consumers Association 2002b)

Labelling seems to have been important in the process of the exit of GM food from retail outlets, but, as will be discussed in Chapter 5 on the EU, the decision to label GM food cannot be put down as the sole reason for this outcome. Its effects were both to enable supermarkets and food suppliers to source non-GM supplies and also as a means of directly empowering consumers to put pressure on supermarkets. I shall discuss the extent to which labelling partly contributed to the political distrust of GM food and the extent to which it causes GM food to be boycotted in Chapter 5.

Food manufacturers and retailers

Although the food industry has had little direct input into the regulation of GM crops (as opposed to GM food), I feel the need to mention the food industry here because they are very influential on issues such as labelling. However, I shall leave the main discussion of their influence to Chapter 5 on the EU as labelling and food safety policy is decided at the EU level.

It is worth pointing out at this stage that the British Retail Consortium has acted in concert with its European peak organisation, EuroCommerce, to support mandatory labelling of GM food through EU law. The food manufacturers, represented in the UK by the Food and Drinks Federation, have accepted mandatory labelling but have been less keen compared to the retailers in extending this concept to derivatives. These include processed oils and starch that were made from GM plant products. Again, this issue will be discussed in Chapter 5.

British retailers and British food manufacturers and retailers certainly supported the introduction of GM food into British markets, and welcomed it at the time as an important technical and cost-cutting innovation. On the other hand the retailers have been pragmatic in reacting to consumer concerns. A tomato puree, made cheaper because of the modification of the tomato for longer shelf life was on sale in the UK from 1996 to 1998, when it was taken off the retail shelves along with other food products labelled as containing GMOs.

Food manufacturers have been dragged along into complying with pressure from retailers to source food products from non-GM sources. Large numbers of consumers actively sought 'non-GM' food. There were some testy moments in relations between European and American food and retail companies as the European companies made demands upon, and successfully persuaded, US food suppliers to source non-GM food.

The UK's Consumer Association has been active in pushing the retail sector to support more comprehensive forms of labelling at an EU level. This has been done in concert with the BEUC, the European Consumers Organisation.

The public

The public is not a network of groups with some similar interests. Yet the networks discussed above try to appeal for the support of the public. This is partly as a means to pressure the politicians who rely on public support to keep them in office. However, it is partly, in this case, because the support of the public through either willingness to consume, or refusal to consume, GM food products has a very instrumental effect on whether the GM crop industry can prosper or even survive.

The GM crop industry had, at the outset, the advantage of the political status quo in which the Ministry for Agriculture, Fisheries and Food actively sponsored their agenda. In addition the food manufacturers and supermarkets saw no technical reason why they should not use and retail GM food products. It was up to the wildlife conservation and radical environmental networks to bring various forms of pressure to bear to win their demands. In fact the public became willing fodder for the prosecution of these demands on account of their fears concerning food safety and their lack of faith in official assurances.

Although the public has shown great hostility to GM food, this hostility is based on grounds that are different to the ideas of the groups in the radical environmental and wildlife conservation networks. Certainly wildlife conservation resonates with the public, but the evidence of opinion research suggests that on its own this would not have been strong enough to generate anything like the degree of public concern about GM food that became evident. For instance, a study on the relationship of the media to public attitudes to risk conducted by analysing data from discussions of panel-groups drawn from the general public concluded that:

> GM food was, perhaps predictably, closely related with recent food scares – in particular BSE (all 16 groups making the link), chemicals in food (7 groups) and salmonella in eggs (6 groups). Indeed in some groups more discussion focused on BSE and beef than upon GM food, partly because people were able to draw upon personal responses to the issue (e.g. beef buying).

(Petts *et al.* 2001: 35)

This conclusion is supported by opinion poll research which suggests that health concerns were suggested by people as potential risks of GM food much more often than other concerns (CropGen 2000). The public focused on the dangers of GM food more than they did on GM crops, according to Eurobarometer opinion research data. In 1999, 53 per cent of the survey said they opposed GM food whereas only 37 per cent said they opposed GM crops (Gaskell and Bauer 2001: 60). Much more will be said about the Eurobarometer data in Chapter 5 when I discuss the EU. However, it is important to attach a note of warning about the use of behaviouralist studies. I have already discussed this in Chapter 2. Here we can see practical examples of how it may be too simplistic to use panel survey results to show that the BSE crisis is the 'cause' of opposition to GM food. How do we know that the BSE crisis is not simply being used as a particularly prominent metaphor for a scepticism of GM food that would be felt regardless of the existence of the BSE crisis? I will debate this issue further in Chapter 5.

Although the anti-GM campaigners did (and continue to) include the food health issue as part of their arguments against GM food it is, arguably, by no means the most important item to the environmentalists. The tabloid coverage of the issue, on the other hand, certainly focused very heavily on the food safety issue. The Daily Mail ran a campaign on the subject in early 1999 with several banner headlines and many other stories containing references to 'Frankenstein Food'. For example, on February 13th the Mail led with the headline 'Frankenstein Food Fiasco'. The article commented that: 'Tony Blair and his ministers' had 'rejected fresh demands for a moratorium on the Frankenstein Foods, despite a stark warning from top scientists that millions could be at greater risk from cancer and infections' (Derbyshire 1999).

Julie Hill, from the Green Alliance, and the green 'representative' on ACRE until its reorganisation in early 1999 commented: 'It was the health editors of the tabloids rather than the environmental groups that focused on the food health issue ... The "Is GM safe to eat?" stuff including how GM food causes meningitis and so on got out of control' (Hill interview – 2001).

There is certainly asymmetry in the attitudes of the public and the environmentalists. Environmental concerns include the intensive agricultural system that GM crops use including pesticide use, the possibilities for 'genetic pollution' to other forms of wildlife, and various impacts on wildlife. The public are much more concerned about the food safety angle. It is, in some ways at least, odd that the wildlife conservation network have appeared to benefit most politically even though their demands are not concerned at all with food safety. I have explained earlier that path dependence can account for the political outcome reflecting the interests of the institution, English Nature, whose work and objectives were already part of government. But this still does not fully explain how the linkage was made between public attitudes which were hostile to GM food and the wildlife conservation network's concerns about the impact of GM crops on biodiversity. So how can we explain this using political science and sociological theory?

Mobilising the public

In Chapter 6 I shall discuss how both Dryzek and the 'difference' democrats criticise Habermas's notion of seeking consensus as the goal of political democracy. Dryzek argues that in practice people come to agreements for different reasons rather than because they have reached a consensus. In the British GM food case it certainly seems that those who are critical of GM crop and food technology can only agree on some policy issues, such as having a moratorium on commercial planting, but their reasons for doing so are different. This, of course, is the key to the answer. The different GM-critical constituencies, the public, the radicals and the wildlife conservation groups were able to unite behind a storyline consisting of a single signifier (word), 'moratorium'. The dominant interpretation of 'moratorium' seemed to shift over time leading to a realignment of forces that wanted to associate themselves with the term.

I discussed Hajer's preferred use of storylines to act as the key pointers to what he called discourse coalitions rather than the interests of the individual groups. This case tends to bear this out since alliances were formed around linguistic devices such as 'moratorium', with changes in their perceived meaning altering the nature of the alliances behind them.

In 1997 Friends of the Earth launched a 'Campaign for Real Food' which featured a drive to 'call for a moratorium (on GM crops) on commercial growing until the risks have been properly evaluated' (Riley and Bebb 1997: 13). This initiative was launched at the time when English Nature were lobbying behind the scenes for research to be carried out on the impacts on wildlife of GM crops before commercial planting on GM crops was begun (Henke 1997).

This apparently united environmentalist public stance continued when, in July 1998, English Nature simultaneously launched a public campaign and sent in a submission to a Select Committee Inquiry into GM crops calling for 'three to five years' research' and 'a period of restraint on commercial releases'. English Nature's public campaign received widespread coverage and hinged on a demand for a three-year 'moratorium' on the commercial planting of GM crops (Vidal 1998a). Nevertheless, the coverage of the demand for a 'moratorium' was still small compared to the coverage that occurred in 1999.

The campaign for some sort of a moratorium on GM crops was the high point of environmentalist unity on GM food and crops. Friends of the Earth began to harden its public interpretation of the term 'moratorium'. In the autumn of 1998 Friends of the Earth and other radical anti-GM food groups started to organise the formation of the 'Five year Freeze Campaign' whose interpretation of the term 'moratorium' involved both a ban on GM food and all open air trials of GM crops, not just commercial crops. English Nature abandoned use of the term moratorium as it was becoming confused with more radical demands and the government drew a distinction between a legally enforceable ban on GM crops which it said would be against EU law, and a voluntary pause. In October 1998 there was a settlement with the wildlife conservation groups which set up the Farm Scale Evaluation trials. The government achieved an agreement with the biotechnology companies for a one-year freeze on commercialisation.

However, these steps only partly satisfied the wildlife conservationists, and in 1999 the issue received intensive media coverage. There does seem to be consistency across papers that I analysed (*Guardian, Independent, Financial Times, Daily Mail*) in that coverage gradually increased in 1998. However, in 1999 there were around ten times the number of stories published about GM food and crops compared to 1998. The event which, above all else, seems to be associated with the biggest leap in coverage was the call by William Hague (the Leader of the Opposition) at Prime Minister's Question Time on February 3rd 1999 for a moratorium on the growing of GM crops. This televised event included Hague telling Tony Blair that 'It is not curious that the Opposition is asking for a moratorium – English Nature is asking for one' (House of Commons Debates 1999). Despite the fact that English Nature was no longer using the term, Hague used the 'moratorium' signifier to mobilise disparate interests to attack the government.

Different projects attempted to use the public's suspicion of GM food for some quite contradictory agendas. Radical environmentalists attempted to frame the issue as one involving 'globalising' multinational corporations conspiring to control farmers and the food supply. On the other hand the right wing *Daily Mail* focused on alleged commercial links between Science Minister David Sainsbury and biotechnological interests and implied that the government's position on GM food was a self-interested Labour conspiracy! All used the 'moratorium' term to advance their projects.

In fact the intense coverage of the GM food and crops issue continued unabated throughout 1999. The government finally concluded an agreement with the biotechnology companies and SCIMAC for a 'voluntary' ban on commercialisation of GM crops in November 1999. This was set to last for around three years until the completion and evaluation of the Farm Scale Evaluation Trials. The number of stories on GM food and crops tailed off in 2000, although there were still more than double the stories compared to 1998.

We shall see in Chapter 5 how the issue of a moratorium on GM food and authorisations featured prominently throughout European discussions of GM crops. An increasing feature of European and UK opposition to GM food and crops has been the emphasis given to the notion of 'contamination'.

'Contamination' is another storyline, albeit a short one that indicates the often quite minimal level of agreement amongst otherwise divergent concerns. First there were the concerns of consumers who wanted to avoid the possibility of being 'poisoned' by GM food and second there were environmentalists who argued that cross contamination, or genetic pollution, was a danger in itself. The use of the notion 'contamination' helped the radical anti-GM environmentalists to brush away the fact that consumers were less opposed to GM crops compared to GM food. If GM crops were an agent of contaminating conventional and organic food with GM food, then they were to be resisted. Hence the anti-GM activists were able to mobilise against trials of GM crops and local media often sported stories of the dangers of contamination from test sites.

The organic farming movement gained considerable publicity from campaigns against nearby test sites, and sales of organic food showed dramatic increases in the 1998–2001 period, although it still claims no more than three per cent of food products, and two-thirds of that is imported into the UK. The Soil Association pointed to research that showed that pollen could be carried for two kilometers or more by bees. Indeed honey producers became so concerned that they sought to ensure that their hives were at least six miles away from GM trial plots. The discursive success of the campaign was such that even officials from the Food Standards Agency would use the term 'contamination' (with its negative associations) when talking about the issue of GM crops cross pollinating with conventional crops (Baynton interview – 2001).

Pro-GM crop interests have often used the 'GM crops are technological progress' storyline as a means of mobilising support, as indicated in the statements quoted in the earlier section on GM networks. Yet this has evidently been unsuccessful in persuading consumers to buy GM foods and in stemming widespread calls for GM crop technology to be put on ice at a political/regulatory level. A problem is that consumers have not noticed, or not been so concerned with, reductions in prices of retail food products and there have not been any increases in qualities demanded by the consumer. I shall discuss the extent to which US consumers give new technologies a greater benefit of the doubt compared to European food consumers in Chapter 4.

The significance and resonance of the 'technology' storyline may be something of an indicator for attitudes to GM food and crops in different places and among different groups. Dobson (1995: 96) has commented that 'suspicion towards technology in general is a fundamental feature of the green intellectual make-up'. Alternatively, to couch it in the way (non-activist) lay-people phrase it sometimes about scientists in the biotechnology or nuclear industries: 'They don't understand what they're doing'. That the scientists and their alleged regulatory scientific friends in government do not understand the ramifications of their activities acts, in the minds of many lay-citizens, as an effective counterbalance to official scientists' claims that the public do not understand rational risk assessment procedures. Herein lies both some of the difference between official scientific and lay-citizen attitudes to risks that I discussed in Chapter 2 and also the linkages between lay-citizen and green attitudes to technology.

It may make sense to argue that the European public is generally more suspicious of some of the more controversial new technologies compared to US citizens. It may be more possible to mobilise Americans by appeals to progress, productivity gains and so on being achieved through technological adaptation.

However, environmentalists are only able to mobilise the public in favour of a green agenda if environmentalist concerns chime with public fears. The public tends to emphasise different points of concern compared to those which are most fondly held by environmentalists. Just as in the case of opposition to waste incineration plant, green groups may be able to gain public support by appeals to a threat to public health (in that case fear of chemicals such as dioxins) rather than appeals to a more ideological green agenda such as supporting waste recycling. In the case of GM food and crops, environmentalists gain the support of the public when they chime with public fears about food safety rather than because of support for the green agenda of shifting towards organic farming practices.

One way of looking at the issue of how different parts of the public sphere become mobilised on this issue is to think of it in terms of the theory of issue framing. I described this in Chapter 2 as the process by which interest groups see a particular issue in terms of how it relates to their perceived interests and way of understanding the world.

A possible way of explaining how it is that a diverse range of interests come to be (coincidentally) critical of GM crop and food technology for a variety of differing reasons is to say that they are all expressing a criticism of the technology through the lens (or frame of reference) of their own prime preoccupations and institutional styles. Hence the radical environmentalists will talk in terms of their opposition to what they see as the established industrial agricultural order and will emphasise the dangers of pollution. Wildlife conservationists will talk about the possible threat to biodiversity, albeit dressed up in the reformist, pragmatic style that they adopt in relation to government, and the public will link the issue to fears about food safety which have been prominent in their minds in recent years. Media will use the issue to promote their own agenda. For example, the *Daily Mail* knows from its readership surveys that a high proportion of its readers are women, and that women are more likely to be concerned about GM food than the men. However, the *Daily Mail* is a pro-Conservative newspaper and so they used the issue to attack the Labour government.

This still leaves the question of what lies 'behind' all these different frames. Is it a matter of general distrust of a new technology that has swept the UK and Europe? I shall discuss the circumstances in which this culture of mistrust has emerged in Chapter 5. However, suffice it to say for now that this distrust is difficult to pin down, and thus, as one member of the UK's 'deliberative' Agriculture and Environment Biotechnology Commission implies, is hellishly difficult to dispel through the sort of positivistic risk assessment in which the scientific and technical elites engage (Grove-White 2001).

Conclusion

I have examined five key parts of the belief system underpinning the UK's system of scientific regulation of GM crops and food which may be useful for comparative purposes with EU and US regulatory practices. First is the notion that GM crops, in common with foreign 'exotic' species, have a unique character compared with conventional crops and indigenous flora and fauna. Second is the attitude to biodiversity issues. This involved an important change in regulatory belief systems which, since 1998, has included the notion that GM crops must not reduce biodiversity compared to conventional crops. This wildlife impact is chiefly construed as meaning the availability of food for birds. Third is the issue of gene flow and superweeds where, despite much debate, the dominant view is still that problems related to gene flow can be dealt with by appropriate crop management. However, demands for 'monitoring' of gene flow are now routinely added as conditions for recommending authorisation of crop trials. Fourth is the issue of antibiotic resistant marker genes. There was division within the regulatory set-up on this issue, but now the dominant discourse rules out use of clinically significant antibiotic resistant marker genes in GMOs. Fifth is the degree of concern about the effect of changes in DNA composition (in GM food compared to conventional variants). The regulatory system now insists on closer examination of what alterations to DNA structure have been made and the possible implications of these changes.

We can link the changes in regulatory discourse with changes in the membership of the Advisory Committee on Releases to the Environment (ACRE). These two elements, discourse and the community of scientists involved in the regulation, comprise what I call the 'discourse-paradigm', the theory of which was outlined in Chapter 2. We can see a change in the policy network concerned with deciding government policy on GM crops which corresponds to the changes in ACRE, the scientific regulatory committee.

In keeping with the approach outline in Chapter 2, I shall use the various dominant discourses of the governing regulatory committees and policy networks (which have been summarised here) as points of comparison with the other case studies, the EU and the USA. However, as discussed in Chapter 2, it is also useful to supplement the discourse –

analytical approach with the notion of path dependence, something borrowed from historical institutionalism. In the case of the UK we can see that the emphasis on wildlife conservation criteria as standards for environmental assessment of GM crops is in keeping with the existing 'insider' dominance of English Nature and the influence of its biodiversity discourse within the government's institutions for environmental protection.

By contrast, the radical environmentalists have little 'insider' presence with government. However, the radical environmentalist general opposition to GM food does resonate with the fears of many consumers who have boycotted GM food products. The anti-GM attitudes of consumers, which are largely related to concerns about food safety, follow on from previous food crises, especially the BSE crisis.

Behaviouralist studies of consumers indicate the importance of the BSE crisis as a trigger of concern about GM food. This is a useful and important finding. Nevertheless, behaviouralism, however important, cannot be taken as a final indication of causation. As has been discussed in this chapter, it is evident that, if we are looking for the causes of differences between UK and USA policy, the BSE crisis is not the only source of difference. The regulatory system is based on different principles in the EU/UK compared to the USA, something that will be covered more in later chapters. It may also be plausible to argue that the way the BSE crisis was interpreted and related to GM food by the public in the UK flows from a greater disposition in the minds of British consumers to be sceptical of certain technologies compared to US citizens.

We can see, through focusing on use of storylines represented by signifiers such as 'moratorium' and 'contamination', how disparate interests are welded together behind apparently coherent projects. By contrast, the pro-GM food and crop interests' use of the 'technology' storyline has had limited effect in advancing their claims. The logic of Dryzek is demonstrated whereby groups reach agreement for different reason rather than consensus as suggested by Habermas. However, if we use Goffman's technique of 'frame analysis' it is plausible to argue that really what we are seeing is that there is a general attitude of scepticism towards GM food and crop technology in the UK which is interpreted by different interests according to their own dominant preconceptions of interests.

It is certainly true that the UK government, as is made clear by various statements made by Tony Blair, does give enthusiastic support to agricultural biotechnology. However, the seemingly polarised debate in the UK should not blind us to the differences that do exist between even the established UK and EU regulation of GM crops and the regulation of GM food and crops in the USA. This should begin to become very clear when we consider the US case study in the next chapter.

Note

1 'Roundup ready' refers to the name of the herbicide, ammonium glyphosate, to which the genetically modified soya bean is tolerant. Roundup ready has been marketed as a broad-ranging herbicide for many years (and before the commercialisation of the GM soya beans), although Monsanto's patent for this herbicide ran out in the year 2000.

4 Forward with technology?

The USA and the GM food debate

Agricultural biotechnology has made enormous impact in the USA, but it is not quite as all-conquering as first impressions may suggest. In 2002 GM crops dominated the US agricultural market in soya and cotton (around 70 per cent) and were building up in canola/oil seed rape (around 55 per cent in North America as a whole), and in corn/maize to around 30 per cent. On the other hand controversy has grown over the 'next' generation of GM crops grown to produce plant-based pharmaceutical products and also over GM animals. Even in the 'conventional' GM sector, development of GM crops has been hamstrung in some areas with major food outlets such as MacDonalds refusing to source GM potatoes. Fears of access to European and other markets being closed has persuaded many farmers and food manufacturers to oppose the introduction of GM wheat, especially in Canada. Even hardened US biotech lobbyists are occasionally wistful about the prospects for agricultural biotechnology. As was stated in Chapter 1, it is also important to remember that only a small proportion, around 6–10 per cent, of the US biotechnology industry has GM crops as its principal business. Even leading biotech companies like Monsanto and Syngenta are, in fact, mainly crop protection (pesticide, fungicide producers etc.) companies, with GM crops as a useful (some would say distracting) complementary product.

As with the case of Britain discussed in Chapter 3, I shall describe and explain regulatory outcomes using the theoretical model developed in Chapter 2. I shall deploy the 'discourse-paradigm approach', which, as I have explained earlier, involves examination of the legal and scientific discourses concerning what are objects of legitimate knowledge, and the political and sociological explanations of focus on these phenomena. This then keys in with the 'discourse analytical' approach that I have also discussed in Chapter 2.

As I did in the case of the UK, I want to begin by looking at the food crisis, or, in the USA, the evident lack of such a perceived crisis. In Chapter 3 I argued that the BSE crisis was a very important factor in fanning fears about GM food technology in the UK. However, the differences between EU and US policies on GM food and crops are not just about the BSE crisis. A different attitude towards food and also towards food regulation may go a long way towards explaining the difference in political and regulatory outcomes on the GM food and crops issue. I shall elaborate on US attitudes to food and then link this to other common US political discourses about the role of regulators and the relationship between industry and regulators. I want to develop this theme as I discuss US attitudes to food and GM food. When I have done this I want to examine key dominant discourses underpinning US regulation of GM food and crops. Then I shall discuss the networks which have been associated with these discourses. First, let us look at how US attitudes to food have formed the backdrop to the debate about agricultural genetic engineering.

Food crisis? What food crisis?

Perhaps I should begin by saying that it is almost axiomatic to this analysis that EU and US consumers have very different attitudes to GM food. This conclusion does not merely rest on my survey of the (many) taxi cab drivers that I encountered during my research trip to the USA, some of whom had never even heard of genetically modified/engineered food, let alone be at all concerned about this innovation. This conclusion rests on hard opinion-survey research which carefully compares US, Canadian and European attitudes to GM food and other biotechnology products. These findings are summarised in Figure 1 below and clearly show that, while US citizens show a moderately positive attitude to GM food, Europeans are quite hostile towards the technology. The Canadians are somewhere in between.

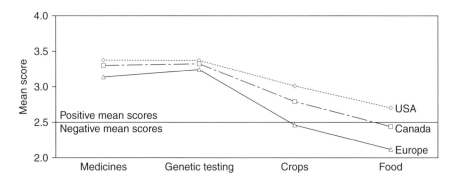

Figure 1 Support for four applications of biotechnology.

Source: Reproduced with permission from Gaskell, G. *et al.* 'Troubled Waters: The Atlantic Divide on Biotechnology Policy', in Gaskell, G. and Bauer, M.W. (eds) *Biotechnology 1996–2000: The Years of Controversy*, p. 108. Copyright 2001 Board of Trustees of the Science Museum.

I must stress that the surveys that produced the results in Figure 1 are commensurable to the extent that the data was compiled using similar questions conducted at broadly similar times and interview conditions. There have been other compilations of opinion surveys which involve comparisons between European and US opinions on GM food and crops in the USA, but I insist that only surveys that use a consistent methodology can be used for these purposes.

Besides the relative (compared to Europeans) lack of concern by the American public towards GM food, two other things struck me about US attitudes. The first was that US citizens, by and large, have a rather different attitude to food in general compared to Europeans, and the second is that the average American seems to have much greater trust in regulators compared to their European counterparts. Let us look at the attitudes of Americans to food. Something that typifies the US attitude to food is to be found in what I felt to be the strange practice, in some Washington, D.C. malls, for food to be sold according to the weight of the food that the customers bought. Fish, chicken, beef, lamb, fries (which became Freedom, not French), vegetables, fruit, cost all the same for each ounce of food that you bought. You just pile it on the plate and the bill is counted according to how much the lot weighs at the check-out counter. This was something I had never encountered before, never in the UK or in the several European countries that I have visited during my years. Never mind the quality, just look at the quantity per buck! (I don't wish to offend, but merely to emphasise the point.)

Some people comment about how the Europeans see food in terms of a quality-type cuisine while Americans see it as fuel, and as a fuel that is best according to its cheapness (Rissler interview – 2003). Hence a much higher premium will be placed on technological means of lowering the cost of production and less on the image of 'naturalness' that seems to appeal to Europeans.

As Krenzler and MacGregor put it:

> In the USA, more than Europe, food has become a matter of efficiency and an expression of technological know-how, being viewed more often as a fuel to be ingested without delay so as to make time for something else. The American 'eating on the move' culture is exemplified by the large number of 'drive thru' fast food outlets in the USA and the fact that automobiles in the USA are sold with cup holders as standard attachments whereas in Europe they are not.
>
> (Krenzler and MacGregor 2000: 304)

Various commentators comment on how, at least until recently, the USA lacked an indentifiable sense of a cuisine of its own. The USA's Greenpeace campaigner on GM food told me:

I'm a cook and I was trained at an American cooking school 20 years ago where all the chefs were European, and when I tell that to my colleagues in Greenpeace Europe they all think that is hilarious – that Americans had to bring over Europeans to teach them how to cook. It's not as true today ... they still have a lot of European chefs, but they also have a lot of American trainers and there are a lot more cooking schools in the USA. I think we're a couple of hundred years behind Europe in terms of really having a food culture we can really call an American food culture.

(Margulis interview – 2003)

I commented in Chapter 3 that while the BSE crisis was a factor in undermining the entry of GM food and crops into European markets in the late 1990s, it would be a gross oversimplification to ascribe this as the principal reason for differences in outcomes in the USA and Europe on the GM crops/food issue. It is not merely that the USA did not have a BSE crisis (in point of fact neither did most parts of Europe itself, apart from the UK) but that the event itself, the BSE crisis, might be interpreted in a different way to the way it was in the UK and the rest of Europe. Positivist theories, for example the advocacy coalition model which I mentioned in Chapter 2, tend to assume that events are interpreted in a set direction. Interpretivist approaches, on the other hand, will focus more on the different ways that events are interpreted or constructed in different contexts by different actors. The point here is that the sense of food crisis which has pervaded European attitudes to food since the 1980s, is not so much in evidence in the USA. This may have nothing to do with events as such, but may have more to do with the way that events are interpreted. It may be that even if there had been a BSE crisis in the USA, it may not have undermined confidence in the safety of GM food in the way that it did in the EU.

This issue of the interpretation of events is also related to the interpretation of uncertainty, which, as discussed in Chapter 2, can be seen as something which is socially negotiated. When the UK government announced, in March 1996, that it had discovered a few cases of new variant CJD among humans, this could have been dismissed (if it had occurred in the USA) as something that would lead to a small, or relatively small number of deaths. As Eric Schlosser, the author of a savage attack on US 'fast food' culture, commented:

As of writing, about a hundred people have died from vCJD, the human form of mad cow disease. Although every one of those deaths was tragic and unnecessary, they must be viewed in a larger perspective. Roughly the same number of people die every day in the United States from automobile accidents – and yet we do not live in fear of cars.

(Schlosser 2002: 284)

We can also see, quite graphically, from this quote that the notion of risk is also socially negotiated. As should be clear from reading accounts, such as made by Jasanoff (1990), concerning the debates about the scientific advice given on occupational and environmental risks, what is at issue is not just the extent of the risk of a particular technology. In addition key issues are, first, what risk from that technology is socially acceptable, and second, what degree of certainty about safety (or, conversely what degree of uncertainty) is permissible before a product or process can be regarded as safe. Indeed, one of my overriding impressions from reading Jasanoff's account is that scientific advisers can only be fully effective in problem-solving when there is consensus about the appropriate level of risk. We have heard about the high risks that people routinely accept with motor vehicles, but mobile phones are another case in point. There have been concerns voiced about the possibility of increased cancer rates associated with mobile phone use, yet this hardly seems to affect people's use of the technology. Very different judgements arise when assessing and managing risks from differing technologies, and it should therefore be no surprise to understand that different cultures may have very different feelings about what is acceptable risk for a particular technology.

It may be that the USA is rather less risk-averse on new food technologies compared to the Europeans. They may be more open to appeals based on lowering food costs through new technology as opposed to appeals based on preserving the 'naturalness' of food. Many in the USA argue that, by comparison with other risks, the Europeans are being irrational with respect to GM food. Are the Europeans too fussy about their food, while the Americans are rational about what they eat? That is a matter of judgement, but given the higher levels of obesity in the USA, the argument that the Europeans are being 'irrational' about their eating choices is unlikely to carry much weight with the average European.

Some might argue that the BSE crisis could never have happened in the USA, or that the Food and Drug Administration (FDA) would have dealt with the matter much more adroitly than the British and European authorities. This is questionable given that the FDA did not move to ban the feeding of various animal proteins to other animals until well after the 1996 BSE crisis in the UK (and well after 1988 when action was taken in the UK). Many argue that the action in the USA is not sufficiently effective even now (Schlosser 2002: 271–289). Certainly, critics such as Schlosser compare the USA's regulatory oversight of food issues unfavourably with Europe (Schlosser 2002: 263–264). Judgements on the relative efficacy of the US and European regulatory systems seem to differ according to their relative support for the food policy status quo. The critics of US food habits and regulation damn US performance and the supporters, such as the American Farm Bureau (which represents US farmers), laud it in comparison to the alleged weaknesses of the European system. As a spokesperson for the American Farm Bureau put it:

I think that's the difference between what you're going through in Europe and what we're experiencing here, in that our government tends to move very, very quickly to remove any suspect product from the supply. And over time, our consumers have felt that they've seen that. Yes, we have recalls, and not necessarily because of GM but because of a variety of different issues. And the public has been comfortable that government moves very rapidly, with the system in place to address those as they arise.

(Watkins interview – 2003)

Are European authorities really complacent compared to US regulators on food safety issues? I am not sure that the evidence supports this assertion. What is more likely to be the case is that the Americans have much greater trust in their regulators compared to the Europeans.

Trust in regulators

American critics of the US food industry complain that the food industry has often been able to manipulate consumers in order to achieve greater freedom from regulation by the FDA (Nestle 2002). Of course, this can be spun in the opposite direction; that US consumers are acting rationally, wishing to lower costs and achieve more product choice through favouring regulatory intervention only when clearly necessary to protect public health. I interviewed a former official of the Environmental Protection Agency (EPA), who now advises companies with biotechnological interests. He supports a 'rational' interpretation of the precautionary principle:

The US National Environmental Policy Act (approved at the end of the 1960s) is a manifestation of what you now call in Europe, the precautionary principle. The idea was if you think you're going to do something that might have significant adverse impact on the environment you must do an assessment before you move forward ... The precautionary principle was never intended in its original formulations in Europe as a means of denying the fruits of technology to society ... To reinvent the precautionary principle to mean that you have to answer every question before you allow a technology to go forward is simply a non-starter ... it will actually have perverse effects on public health and safety.

(Abramson interview – 2003)

The incidence of contamination like *E coli*, salmonella, listeria and so on is certainly no less severe in the USA compared to the EU, yet faith in the USA's regulatory procedures does not seem to have been seriously dented. Incidents such as the outbreak of eosinophilia myalgia syndrome which is said to have killed 37 people and disabled a further 1,500 in 1989 do not seem to cast doubt on the efficacy of the regulators themselves. I mention this particular incident (known as the L Tryptophan affair after the name of the food supplement) partly because it was associated with a genetically engineered product. The industrial process has been blamed rather than the GM enzyme, but one could expect a different interpretation of such an affair in the EU. When problems do occur in the USA, and there have been some more recent controversies over GM food (as I shall discuss soon), a common discourse seems to be that the 'prompt' action taken by the FDA to deal with problems confirms the ability of the FDA and other regulatory bodies to defend the health and interests of the consumer.

Indeed, several US states have libel laws aimed specifically at stopping people implying that certain types of foodstuffs are unsafe to eat. Oprah Winfrey found this out when she criticised beef consumption, and this in a country that places a high premium on freedom of speech. She said that she would not eat another burger on account of mad cow disease risks. In fact the courts eventually ruled that Winfrey had not infringed Texan food disparagement laws. However, in the UK, where libel laws are generally a lot stricter than in the USA, it would be necessary for someone to libel a specific company (rather than a general foodstuff) before an action could be successful. In the USA the food producers appear to have special protection in many states.

Some would argue that this state of affairs is proof of corporate conspiracy theories, with the big corporations manipulating decisions behind the scenes, and bribing or rubbing out potential opponents. Maybe they do! One could well construct the USA as run by the corporations, but if so it may be because American voters are not too concerned that this is the case! This brings me on to a second key difference between European and US attitudes, and that is trust in the regulators.

Marcia Mulkey, a key official of the Environmental Protection Agency dealing with GM crops, has captured the mood of apparent trust in regulators. In welcoming the contribution of scientific advisors at the beginning of a Scientific Advisory Panel (SAP) public hearing about a new type of Bt corn crop she said:

> All of that contribution that you add to what we do is valued by us and, I believe, valued by our public. And it is never more obvious than in this subject matter involving genetic modification that the American people have a degree of trust in their government around these issues, which is not enjoyed in every part of the world on topics close to these.

> (FIFRA Scientific Advisory Panel – 2002)

This interpretation is widespread. Opinion surveys have indicated increasing trust in regulators in the USA in the 1990s (Etzioni 1996 cited by

Lofstedt and Vogel 2001: 402). Even those who regularly organise anti-GM lawsuits recognise the considerable prestige of the FDA among voters. For example, a leading legal advocate at the Center for Food Safety told me:

> In the early 1990s the FDA was led by a dynamic guy called Michael Kessler who took on tobacco as a health issue. He got a reputation as a champion of people's rights against being killed by second hand tobacco smoke. So the public persona of the FDA is very positive.
>
> (Mendelson interview – 2003)

Of course the FDA has a very long pedigree. It was established in 1927 whereas European food agencies and environmental agencies have only been established on the basis of (and no doubt in response to) the US model in recent years. On the other hand, as analysts of regulatory politics assert, the tradition built up from the inception of the Environmental Protection Agency at the beginning of the 1970s was that regulators were mistrusted in the USA, and challenged in court, whereas European regulators built up trust with the industries that they regulated through consensus. Some analysts believe that the picture had changed by the 1990s, with European regulators becoming mistrusted and there being a move to a more US-style of system where risk assessment was separated from risk management (Vogel 2001; Lofstedt and Vogel 2001).

I do not doubt the accuracy of this analysis in terms of how people see current realities. However, the events (e.g. the food crises) that they hold to be influential are themselves constructions which are capable of various interpretations, such as whether the regulatory authorities acted effectively. Generally speaking it seems that the more radical the dominant public attitude on environmental issues becomes, then the less likely it is that the regulators will be given a good press. So why is there this apparent difference between trust in US and EU health and environmental regulators today? I argue that an explanation for this rests on sociological factors concerned with the relative strength and patterns of radical environmentalist opinion and activity. This is as opposed (simply) to different events promoting differing public attitudes to regulators.

Observers may cite particular events for generating confidence in regulators, such as the FDA campaign over smoking, but analysts have pointed to a more general trend towards more pro-industry regulation in the USA compared to the EU on issues ranging from GM food, considered here, to plasticisers in children's toys and mobile phones (Lofstedt and Vogel 2001: 403–404). The explanation for US public confidence in the regulators is not the stringency of the actual regulatory regime which (especially in the GM food and crops case) may nowadays be more permissive towards industry in the USA compared to the EU. The current trust in US environmental and health regulators is a change from the situation in the late 1960s and 1970s, and particularly a change from the situation when the EPA was established in 1970 on the top of a surge of environmentalist social concern. In the late 1960s and early 1970s US environmental regulators seemed to be mistrusted in the USA more than they were in Europe. As Chris Rootes, a leading sociological analyst of environmental movements puts it:

> At that time (the end of the 1960s) the USA was a very divided country with tens of thousands being killed in the paddy fields of Vietnam. You've got to remember that environmentalism had also been very strong since Rachel Carson's campaign against pesticides started with her book in 1962. Environmentalism was seen as a means of unifying an otherwise divided country, and Nixon took advantage of that with the establishment of the Environmental Protection Agency.
>
> (Rootes interview – 2003)

Great social divisions and political activism will often generate new forms of activity as the 'politically trained' activists move on from the original focus of concern. As a researcher at the American Enterprise Institute has commented:

> I remember working as a Congressional aide in the late sixties. Across the street from our office was a building – the Methodist Building – that was filled with antiwar activists. And I remember thinking, 'What are these people going to do when the war is over?' They had learned a skill – political organizing – and they proceeded to put it to use. There was an explosion of 'public interest' activist groups in the early seventies, and they had a significant impact not just on political reforms, but also on environmental and consumer regulations. Then, in the late seventies, there was an explosion of industry groups and trade associations to counter-act them.
>
> (Norman Orrnstein quoted by Klein, 2002: 95–96)

Hence social attitudes towards food safety may have been much more permissive towards industry in the USA compared to the EU long before the BSE crisis. It is better to understand the differences by regarding 'food crisis' events, to a large part, as have been socially constructed rather than regarding different social attitudes having been simply created by different food events in the USA compared to the EU.

However, I also believe there is good evidence for a storyline that, on a more general level, US environmental public policy out-turns were much more radical than the EU's in the 1960s and 1970s, but weaker since the 1980s. This becomes more apparent when you trace policy outcomes across these periods. Every analyst of international environmental policy worth his/her salt knows that it was the USA that took the lead in proscribing use of ozone-eating chlorofluorocarbons (CFCs) in the 1970s while the EU lagged behind the USA in taking action until the end of the 1980s. Unleaded petrol was on sale in the USA in 1971, many years before most of Europe and clean air acts were on the statute books long before Europe. In the 1970s the US environmental movement was noted for its vigorous pursuit of its agenda through the courts which themselves developed their own tendency for judicial intervention to protect the environment (Jasanoff 1990: 49). Certainly, for my part, when I was writing my first book (about energy and environmental policy) at the end of the 1980s I naturally took most of my examples of good practice in energy efficiency and renewable energy from the USA. Although there is still some good practice going on in the USA, I am (when writing about such matters today) much more likely to turn to places like Denmark, The Netherlands or Germany for green energy practices that developed in the 1990s.

Some have highlighted the ALAR controversy, in 1989, when the EPA stepped in to ban a pesticide used on Californian apples, as an example of how environmental regulation has been 'strict' in the USA. However, the very fact that this has gone down in the lexicon of US chemical industry lobbying as the prime example of regulatory overkill can also be taken as an example, indeed an exemplar, of how regulators have (since) been persuaded to be more sensitive to industry in their mode of regulation. This particular decision has been constructed as a ban too far. However, the EPA's stand is still hotly supported by environmentalists who argue that it was secondary products of the pesticide that presented health risks rather than the product itself. Indeed, the leading environmental campaigner against ALAR argues that the enactment, in several US states, of laws banning criticisms of particular types of foodstuffs (for example beef or chicken) is a direct consequence of the 'regulatory overkill' construction of the ALAR case. It is argued that this interpretation has been successfully constructed to advance a (pro-industry) discourse that irresponsible scaremongering by activists has harmed industry and has raised food costs (Negin 1996).

Some would contest the view that environmental regulation has become stricter and Europeans more risk-averse compared to the USA (Rogers 2001). Certainly it is reasonable to assert that action against passive smoking has been farther advanced in the USA compared to many parts of the EU. The USA is preventing stem cell research while it is permitted in some EU countries. However while these sorts of issues are of wide-ranging societal concern, these types of issue are not specific campaigning priorities of environmental groups. Certainly the difference on stem cell research owes a lot to the strength of religious groups in US politics, and is most certainly not a product of green campaigning.

I shall comment more in Chapter 5 on environmentalism in Europe, but suffice it to say that European environmentalist mobilisation was (generally speaking) later in developing compared to the USA. It began to come of age in Europe in the early 1980s in terms of campaigns on acid rain and the electoral rise of the Green Party in West Germany. By the end of the 1980s a strong environmental surge had spread across Europe, with global issues such as ozone depletion and global warming rising up political agendas. It was in the 1980s that regulation on GM crops was formulated, and it should be noted that the GM issue arose at a time when US environmental policy was becoming relatively more conservative (especially on issues which lacked the inertial push of the 1970s) while European policy was becoming more radical on environmental issues.

I think it useful to mention a point here that I wish to develop further later: namely that what is important in the study of environmental regulation is not so much whether there is a clear division between risk assessment and risk management, for I doubt that it is this that really makes a difference in policy out-turns between US and EU regulatory models. I would argue that political factors, that is the nature of the dominant social discourses underpinning regulatory, quasi-scientific paradigms, will set the tone for both risk assessment and risk management decisions. I regard it as a nonsense to try and separate out risk assessment from risk management anyway. Both are guided by dominant political discourses of the day and also the ways in which events are constructed.

Nevertheless, it would be very wrong to picture governmental decisions about GM food and crops in the USA as representing perpetual calm and total agreement with the desires of the agricultural biotechnology industry. Some change in attitudes can be gleaned by the fact that since the year 2000 the regulatory authorities have felt compelled to act to deal with what could be termed 'GM food crises' in the USA. I want to have a brief look at these.

GM food crises?

The biotechnology and the majority of the farming and food processing industries would probably demur from saying that there have been GM food crises, but the 'Starlink' and 'ProdiGene' affairs have made people nervous in the food manufacturing industry about consumer reactions. They have also given the critics of GM food something to campaign about. First, the 'Starlink' affair, which has been the most serious GM food 'crisis' in the USA to date.

In 1997 the Environmental Protection Agency allowed the limited registration of a new type of Bt corn called 'Starlink'. Like other types of Bt crop, the corn produced its own insecticide in its pollen to protect the crop from pests (quite a wide range in this case). However the toxin produced by the modified plant was suspected to be similar to proteins that produced allergenic reactions in humans. Hence the EPA limited its registration to non-human food uses, mainly animal feed.

Unfortunately there were reports that there was widespread 'commingling' with regular corn used in human food throughout the harvesting, transportation, distribution and processing stages. The reports trickled through to Friends of the Earth who commissioned tests on taco shells. Traces of the Starlink corn were found in September 2000. The FDA confirmed these tests the following month and traces of Starlink corn were found in a number of other foods, including food exports. EPA registration of the corn was withdrawn and a number of companies withdrew stocks of foodstuffs from retail outlets. The affair triggered a wave of publicity, although what is also notable about this affair is how the disclosure of this 'crisis' did not lead to much of a consumer reaction, like the shoppers panic that would surely have occurred in Europe! As the *New Scientist* magazine commented:

> Last month, American consumers learned that they might have eaten taco shells containing genetically modified maize that had not been approved for human consumption. Did they boycott Mexican food? Did they picket supermarkets? No. While a few green pressure groups tried to turn the drama into a crisis, indifference ruled from coast to coast ... These events reveal the transatlantic gulf in attitudes to genetically modified foods.
>
> (*New Scientist* Editorial 2000)

The language used was also different. In the EU when GM crops or food that have even been approved for human consumption are mixed in with conventional or organic crops it is called 'contamination', not 'commingling'. Indeed the 'commingling' term was used in the Starlink affair even by Marion Nestle, someone who is not generally thought to be an ally of agricultural biotechnology companies (Nestle 2003: 12). Of course Europe has seen sustained consumer boycotts of legal, never mind unregistered, GM food crops. In the USA the retail outlets are likely to react more on the fear of being sued by people alleging allergic responses rather than by fear of boycotts or consumer resistance. This affair reveals not only a different set of reactions to GM on different sides of the Atlantic but also a different sensistivity to alleged food crises themselves. An event labelled as a food crisis in the EU may be constructed as something less of a crisis in the USA. This is another example of how 'positivist' analyses, which stress how events change belief systems, may sometimes deliver simplistic analysis because the same event can be constructed in different ways. Yes, we can suggest that events change belief systems, but we also need to consider how things work the other way around; that is how belief systems condition the way people interpret events.

A more recent 'crisis' or 'regulatory problem' occurred in the shape of the 'ProdiGene' affair. ProdiGene is the name of a biotechnology company that has been researching and testing the use of genetically modified food crops to produce pharmaceutical products. These are called pharm-crops or plant-based pharmaceuticals. ProdiGene tested production of a corn modified to produce a pig vaccine to treat pigs for swine fever. Fields in sites in Iowa and Nebraska were sown with the pig-vaccine corn. However in November 2002 the FDA ordered the destruction of several thousand bushels of corn which had (as Europeans would put it) been 'contaminated' with the altered pig-vaccine corn. Although there is no evidence that ingestion of small amounts of pig vaccine will harm humans, it had not been approved for human consumption by the FDA. The FDA also ordered the destruction of half a million bushels of soya beans which had been harvested on ground which had been used to grow the GM corn in the previous growing season. Corn leaves and stalks had become mixed with the soya. Many in the food industry felt that whatever the safety issues may be, this was a potentially big turn-off for consumers. People who want to be immunised against swine fever via their cornflakes would presumably ask for it!

What heightened the political significance of this affair was that ProdiGene had been a vocal campaigner against regulations involving lengthy separation distances from more conventional crops. The upshot was a voluntary agreement by the Biotechnology Industry Organisation for a moratorium on pharm-crop testing in the Corn Belt (Gillis 2002). Food industry organisations such as the Grocery Manufacturers of America have been demanding stricter controls on pharm-crops, something which I shall discuss later.

Indeed food companies and consumer organisations are also sceptical of another emerging biotechnological innovation, that of GM animals. In February 2003 the FDA announced an investigation into fears that 386 genetically engineered pigs had been sent to livestock markets by University of Illinois researchers. The pigs had been genetically engineered so that some levels of growth proteins were enhanced.

We can see how US attitudes to GM food have developed from a context of what could be said to be a more 'pro-industry' attitude to GM food and food quality in general compared to the EU. In parallel with this the GM food and crop regulatory system appears to be less restrictive than the EU's regime. I want now to explain and analyse this regulation in order to focus on some key discursive points which can be compared to the UK and the EU. Let us begin with the genesis of the GM food and crop regulation in the USA. This was formalised into the 'Coordinated Framework'.

The Coordinated Framework

Although the 'Coordinated Framework for the Regulation of Biotechnology' (Office of Science and Technology Policy 1986) was promulgated as long ago as 1986, it serves, virtually unchanged, as the basis for regulating biotechnology in the USA today. The whole thrust of the 'Coordinated Framework' stands as a monument to the differences compared to the approach which was formally adopted by the EU just four years later. The two systems have drawn even farther apart since then with the US system remaining virtually unchanged and the EU's framework for biotech regulation becoming more restrictive.

Early experimentation with GM crops had already begun by the 1980s, and in 1983 a major controversy erupted regarding the proposed release, by a company in Oakland, California, of some GM frost-resistant bacteria. This was intended to help crops such as strawberries survive cold weather. Jeremy Rifkin and his Foundation for Economic Trends succeeded in what is still their only court victory in blocking GM technology. They delayed, for four years (starting in 1983), efforts to test the technique in an open-air trial.

This concern formed part of the backdrop to the deliberations (in 1984), by a group of cabinet and senior food, environmental and science policy officials, to set up a policy on regulation of biotechnology products. The proposals were published through the White House's Office of Science and Technology Policy (OSTP). The other part of the backdrop was a strong deregulatory atmosphere pervading the Reagan administration. Research on public opinion suggests that support for more environmental regulation was generally at a low ebb in the USA at the beginning of the 1980s, although attitudes became more environmentally concerned as the decade moved on (Dunlop 1991). There was certainly no doubt about the prevalent attitude at the White House and among many administration officials.

Charles (2001) charts the story of how Monsanto's chief negotiator on the new biotech regulations (Will Carpenter) actually had to fight Reagan Administration officials who wanted no regulation of GM crops and food at all. Carpenter had been a warrior in the battles with environmentalists a few years earlier, and he wanted a consensus approach to regulation rather than the confrontation that had characterised many arguments in the 1970s. Carpenter was amazed at being called a traitor and an ally of environmentalist 'troglodytes' by Henry Miller, the FDA biotechnology spokesman (Charles 2001: 28). In fact Monsanto (allied to the EPA) largely got its way on the regulatory approach. This involved the notion that a specific framework for regulating GMOs should be agreed by the government, but one which was based on existing laws and one which developed existing regulation and assessment of conventional food and crops. The Coordinated Framework document cited a contemporary OECD Report (1983) as providing the justification for, and basis of, this philosophy. It summarised the OECD as saying:

> The means for assessing rDNA organisms can be approached by analogy with the existing data base gained from the extensive use of

traditionally modified organisms in agriculture and the environment generally . . . There is no scientific basis for specific legislation for the implementation of rDNA technology and applications. Member countries should examine their existing oversight and review mechanisms to ensure that adequate review and control may be applied while avoiding any undue burdens that may hamper technological developments in this field.

(Office of Science and Technology Policy 1986: 14)

One can see from this excerpt that the Coordinated Framework implicitly accepted the notion, supported by the biotechnology industry (but opposed by GM sceptics and anti-GM activists), that GM agriculture was merely a development of, rather than a break from, conventional breeding techniques. Consequently the new technology could be regulated under existing laws. GM crop and food innovations could be regulated as 'products' rather than treated differently to other crops and food as a different 'process'. Indeed the Coordinated framework stated:

Upon examination of the existing laws available for the regulation of products developed by traditional genetic manipulation techniques, the working group concluded that, for the most part, these laws as currently implemented would address regulatory needs adequately. For certain microbial products, however, additional regulatory requirements, available under existing statutory authority, needed to be established.

(Office of Science and Technology Policy 1986: 3)

This seems to suggest that if it had not been for the concern generated by the campaigns of Rifkin and others about the frost-resistant bacteria release (mentioned earlier) there would not even have been any distinctive regulation of GMO releases at all. The chief concern of the Coordinated Framework and latter-day reviews such as that of the National Academy of Sciences (NAS) (2002a, 2002b) has been to set out how the FDA, the EPA and the USDA could work together effectively to regulate individual applications of GM food and crop technology. Although this has been done by basing the issue of rules and policy advice on existing laws, sometimes the existing law has been interpreted by the Coordinated Framework in a rather creative way.

The United States Department of Agriculture's (USDA's) role in GM crop regulation is carried out under Plant Pest legislation which concerns 'movement into or within the United States of organisms that may endanger plant life and to prevent the introduction, dissemination or establishment of such organisms' (Abramson and Carrato 2001: 247). This has been taken to cover GM crops since their process of modification usually involves (see Chapter 1) use of *agrobacterium tumifaciens* as a carrier for the genes which are transferred. In fact, not all GM crops are generated this way, but crops produced in other ways have nevertheless been treated in the same way.

Herbicide tolerant GM crops are administered by the USDA as plants. Their tolerance of herbicides will hardly feature in their assessment at all since the herbicides involved are already registered for use with the EPA. The notification of new use has not involved extensive assessment of environmental or health implications. Moreover there have been no British-style controversies about indirect effects of herbicide regimes on wildlife. In the USA, bird protection groups tend to be more concerned about the fate of rare birds than the impact of agricultural practices on common bird populations. Unlike the UK and much of Europe, only a minority of the US countryside is farmed.

There have been controversies about the effects of Bt crops on non-target insects. Bt (pesticide-producing) crops are regulated by the EPA under the Federal Insecticide, Fungicide and Rodenticide Act (FIFRA). In fact the principle concern of the EPA has been to adjudicate about arguments over whether insects will develop resistance to the Bt toxin. Since 1999, the EPA has given attention to concerns that the pesticidal properties may harm non-target insects such as the monarch butterfly. The EPA also has authority under the Toxic Substances Control Act (TSCA). This was given rather more importance under the Coordinated Framework than it has received in practice. Since the 1980s the focus of agricultural biotechnology innovation has shifted from spraying crops with GM microbes (which stirred up a great deal of controversy) towards modifying the crops themselves. There has been a decline in commercial interest in this form of GM agriculture.

GM foods (as distinct from GM crops) are regulated, in a formal sense, under the Federal Food, Drug and Cosmetic Act (FFDCA). In a real sense they have not been much regulated under this Act at all, since only GM foods that are regarded as being not 'substantially equivalent' have to undergo a mandatory scientific assessment. Most GM foods are regarded as being the same, for food purposes, as other foods derived from the same, albeit unmodified, crop.

Underpinning this regulatory approach is the notion of avoiding undue burdens for technological advance. This is a consistent (and compared to the EU rather different) theme in US regulation of agricultural biotechnology. Even in the more GM-precautionary atmosphere after the turn of the century the National Academy of Sciences placed considerable emphasis on reducing regulation where it was not needed and also on reducing the burdens of regulation on small biotechnology companies (National Academy of Sciences 2002a: 179–181). This emphasis on minimising regulation is a key facet of the free market discourse which has dominated the USA even more thoroughly than the EU (in GM crop policy, for example), since the Reagan years.

A free market approach

Val Giddings of the Biotechnology Industry Organisation succinctly characterised the general 'free market' approach when he commented (on the subject of regulation of GM crops):

> In the USA it is accepted that it is not the Government's responsibility to regulate the market place. Its duty is only to ensure safety. So permission to conduct commercial activity is not required. Safety is the key consideration, and that is what is regulated.
>
> (Giddings interview – 2003)

Some would argue that this 'free market' conception emphasises a 'free market' for corporate producers. This 'free market', rather than what could be called a more 'interventionist' style is, I feel, a more relevant way of contrasting the US and EU approaches to regulation of GM crops rather than the more traditional comparison between 'adversarial' US and 'consensus' EU styles. This traditional approach is articulated by Levidow and Carr, when they say that:

> US regulation presents a formalistic rule-bound image. By contrast, for example, regulators in the UK regard such formal scientific method as mere 'competence'; their true expertise entails an informal negotiation which builds trust with those being regulated ... US risk regulation has been theorized as 'adversarial'. Regulators are expected to devise detailed rules, to justify them in clear scientific terms, and to enforce compliance. The scientific basis of risk assessment readily comes under challenge from outsiders such as industrialists or environmentalists.
>
> (Levidow and Carr 2000: 11)

This concentration on formal procedure may have helped understand differences in the 1970s when an apparently more adversarial US regulatory formal style seemed to match the rather more conflictual public tenor of environmental politics in the USA.

Indeed it is the case that the US system has a much greater degree of public consultation built into the system of risk assessment and risk management compared to the elitism of the British and EU systems. However, this facet is of less importance in this comparison of GM agricultural regulation simply because the US public has not done a tremendous amount to take advantage of its rights to engage in debate with regulators! At least this was the case until 1999 when the GM crop and food industry had become well established.

In the USA there is a statutory period for public comment whenever an Agency like the EPA proposes a new rule, or when a new, say GM crop, product is proposed for registration as a 'de-regulated' product (hence cleared for commercialisation). By contrast, as we have seen in the UK, regulatory judgements are made by scientists making recommendations to ministers in accordance with well-defined, politically based, criteria for assessment. Up to now there has been no formal role for public comment in this process, however much public comment there may be occurring outside this formal system. There is more public information in the EU on some matters. For example, in the EU there is much more information about where GM crop trials are occurring. In the USA the locations of trial sites are closely guarded secrets, whereas in the EU there have been irresistible calls for trial sites to be publicly listed.

Certainly the contrast between an elitist European and a more adversarial US style does not help understand the comparison between US and EU *outcomes* in GM food and crop regulation. Arguably it does not help us understand a more general comparison of US and EU outcomes in other areas of environmental regulation since the 1980s. This is because in the EU the 'elitist' industrial scientists appear to have lost out more often in the regulatory battles compared to industry in the USA. Here there have been accusations of 'revolving doors' between regulators and consultants for industrial interest groups.

An effective way of comparing outcomes is to compare the discourses that underpin the regulatory decisions. A vitally important discourse is the 'free market' discourse of environmental regulation in the USA which I have earlier described. The free market approach is not just a style of regulation, it is also a dominant political discourse. It lends itself to the close relationship between the industrial interest groups and the regulatory apparatus that has been a much observed (and propagandised) aspect of US GM food and crop regulation. Given the degree of policy accord that has existed between US Administration officials and the biotechnology companies since the 1980s (even to the extent that the biotech companies actually allied with the EPA against those who wanted less regulation) it is hardly surprising that there are 'revolving doors' between the officials and biotechnology interest groups. Officials would have to change none of their views to move from being a regulator to being an industrial lobbyist. Indeed EPA officials firmly believed (and generally still do believe) that GM crops represent(ed) environmental improvements, not threats.

The point I lead onto here is that we are going to understand the differences between GM regulation in the USA, UK and EU by analysing the discourses that underpin the scientific-regulatory apparatus. I do this according to the 'discourse-paradigm' approach that I developed in Chapter 2 and have deployed with effect in the UK case study in Chapter 3. Indeed I want to compare the discourses on many of the same topics that I focused upon in the case of the UK.

I am not going to argue that these discourses 'cause' events. Rather, as the discourse theorists put it, they 'constitute' reality. It is wrong to understand the differences between UK, US and EU GM food and crop regulation as being caused by different sets of regulations and rules, a formulation that may be preferred by rational choice institutionalists. I am not going to be dogmatic and insist that rules never lay down incentives and have clear effects that can be traced directly to those incentives. However, on the question of whether GM crops and food should be given the benefit of the doubt, and industry left to get on with things, US opinion has generally adopted a free market discourse which is a cognitive, belief, structure system rather than an incentive structure.

It would, of course, be wrong to conceptualise all of US thinking as being entirely dominated by the 'free market' 'leave industry to get on with it' approach. There have been elements supporting a more European, interventionist approach, but they have always been outsiders. Although the battle in government circles (at the time the Coordinated Framework was being formulated) was between a virtual free market and minimalist regulation for the purpose of boosting consumer confidence, outside government there was an environmentalist third force. They advanced what was to become the dominant discourse in the EU, that is the need for special legislation. The Union of Concerned Scientists (UCS) lobbied for special legislation to regulate biotechnology, and in particular to regulate agricultural biotechnology as a separate process rather than merely on a product-by-product basis. The UCS analyst Margaret Mellon called on Congress to 'enact a completely new statute establishing a comprehensive program for the regulation of the environmental release of genetically engineered organisms' (Mellon 1988: 49) It is a call that has never received serious attention either in the White House or on Capitol Hill. As Jasanoff (1995) commented:

> Confusion in regulatory circles, and associated boundary disputes over expertise and authority, rekindled interest in a legislative solution to managing biotechnology, but political pressure was insufficient to overcome a settled congressional reluctance to do anything that might endanger the US industry's competitive position ... The courts, which in the American political context might have provided an independent spur to a broader public debate on biotechnology, proved unusually quiescent throughout the period of development.
>
> (Jasanoff 1995: 317)

Mixing science and politics

In the case of the US system, the members of Scientific Advisory Panels (SAPs) are drawn from the leading scientific associations. The EPA, the FDA and the USDA all have their SAPs, although the USDA has relied heavily for its advice on the NAS which it commissioned to produce a range of reports in the 1980s and 1990s. Much of the advisory work conducted for the USDA is concerned with receiving advice from representatives of relevant interests rather than scientific advice per se. By contrast, the EPA's SAPs have a long pedigree. The EPA's discussions on GM crops are organised through its FIFRA (Federal Insecticide Fungicide and Rodenticide Act) SAP. This includes seven permanent members who are appointed by the EPA on the basis of nominations submitted by the National Science Foundation and the National Institutes for Health. According to an EPA official:

> EPA may augment the permanent membership with experts chosen to particularly address that subject. Those experts who are added to the panel are identified from a number of sources, including EPA's own familiarity with the leading researchers in the field, recommendations from the range of stakeholder groups including industry and environmental advocacy groups, and nominations from the public in response to a general notice and call for nominations.
>
> (Interview with EPA officials – 2003; names removed to preserve anonymity of officials)

Analysis of membership of SAPs concerned with recent issues indicates that while the SAPs consisted of academics entirely relevant to the issue under discussion, the topics selected for discussion were clearly those that reflected political concerns of the day regarding GM crops. For example, the SAP on 'Bt Plant-Pesticides Risk and Benefit Assessments' which deliberated in October 2000 comprised (in addition to EPA Administrators) eight entomologists, four concerned with agriculture or insect issues in agriculture, one economist, one environmental toxicologist, one environmental biologist and one applied economist (FIFRA Scientific Advisory Panel 2000c). This was clearly in response to the emergence of impact of Bt plants on non-target insects in the wake of the studies suggesting that Bt toxins harmed monarch butterflies.

The Food Advisory Committees which advise the FDA are slightly different in that they will often involve consumer and special interest representatives to join in the discussions with scientists. Sometimes (for example to consider the implications of the Starlink affair) hybrid panels will be formed including scientists nominated by different Agencies. An EPA official described the format of a panel deliberation as follows:

> We announce the meeting, we make everything that the panel members are going to consider available to the public, we invite the public to comment in advance of the meeting, EPA develops questions to the panel members on the material that we have provided the panel members with. At the public meeting, EPA makes presentations. The panel members ask questions, then the public is invited to offer their comments oral presentations or slide shows, then the panel

deliberates in public, and at the end of the panel deliberation they produce a written report which is made public and is sent to EPA as their response to the questions that EPA has asked.

(Interview with EPA officials – 2003; names removed to preserve anonymity of officials)

Officials (and the dominant interest groups) insist that their decisions are based on a 'science based' approach. Indeed this ascription 'science based' is often produced when comparing US regulation with the alleged inadequacies of the EU's approach. Nevertheless, US regulatory decisions on environmental and health issues are dependent on political judgement both within and beyond the EPA, FDA or USDA. Agencies must, among other considerations, worry about possible legal challenges. The courts themselves make judgements based upon the perceived social truths of the day, and these comprise dominant discourses in which both scientists and lay people are routinely immersed.

An EPA official described the decision-making system as such:

The Administrator (the EPA's Boss), under the law, has the authority to make decisions to execute the law and the Administrator may delegate that authority as far down into the organisation as she chooses. In the case of registration decisions, that authority is delegated to the Director of the Biopesticide and Pollution Prevention Division ... But because of the significance of these decisions, and because (the Director) is a wise and politically savvy person, (s)he pays attention to his/her boss and his/her boss's boss all up to the Administrator, if the decision is one that is controversial enough or significant enough to warrant their attention ... The reality is that these decisions are made through a mix of staff work and careful consideration of legal and policy considerations that are brought in at different levels.

(Interview with EPA officials – 2003; names removed to preserve anonymity of officials)

None of this implies that the deliberations of the SAPs are unimportant. As Jasanoff (1990) has discussed, they act as interpreters who convert arguments about values into scientific tests and technical language. They are a medium of exchange whereby the legitimacy of the decision-making process is bolstered. However, the scientific advisers cannot be divorced from political influence for three reasons. First because, as I have just stated, they are influenced by whatever social truths of the day are prevalent among the population, for example, an American attitude that has not been as sensitised to food safety risks as the Europeans and also a set of dominant food interest groups that protect the biotechnology industry. Second, the membership of SAPs tend to comprise people who enjoy a culture of working with the EPA and other agencies (Jasanoff 1990:

93–95). Third, the issues that are selected by the EPA etc., and the questions that are asked of the SAPs form the scope of the possible decisions that can be taken.

Because the formal decision-making process is based around Administration officials rather than the SAPs my principle source of texts for analysis of key discourses are the policies and rules promulgated by the Administration, as well as the Coordinated Framework document which I have already utilised. There have been occasions where the Agencies have not followed the advice of the SAPs, but what does seem apparent is that the SAPs have shared the essential nature of the dominant discourses underpinning US GM food and crop regulation. An EPA official commented to me that there has been a tendency towards split decisions on the subject of biotechnology, but the only occasions where the EPA does not take the advice of the SAP are on research priorities. As an official put it:

> Academics can come up with all kinds of interesting things but maybe in the real world you can't actually apply that or it wouldn't really change the risks ... An example of that one is the reassessment of all the BT crops that were there finishing in October 2001. We had a group of scientists on that SAP who believed that we ought to go in and know everything that was going on in the soil microflora when you plant BT crops. And we had other scientists who said 'you don't know enough about the soil microflora to make this make any difference at all' ... We agreed with the group that said 'what are you going to do with this information? You'll cost huge multimillions of dollars even trying to answer this question and what would you do with it in the end?'.
>
> (Interview with EPA officials – 2003; names removed to preserve anonymity of officials)

Efforts have been made in recent years to react to criticisms about commercial links between scientific advisors and industry. EPA conditions have been tightened so that since the summer of 2002 scientists will be debarred from serving on SAPs if they have grants or contracts with biotechnology or other relevant industrial concerns. Previously the 'interest forms' which aspirant SAP members had to complete only asked about ownership of commercial stocks and positions as employees.

I discussed in Chapter 3 how, in the UK GM crop regulatory system, particular scientific-positivistic tests were selected to generate knowledge, and therefore truths, not because they represented the discovery of some absolute reality, but because such tests were authorised by the sets of values held dear by the ruling policy network. I argue that the US system can be understood in exactly the same way, even though there is a theoretical division between risk assessment conducted by Scientific Advisory

Panels (SAPs) and risk management conducted by officials of the EPA, FDA and USDA. Decisions are political, and while the advice of SAPs occasionally deviates from what the Administration officials decide, this deviation is slight compared to the agreement.

The choices about what issues are discussed by SAPs are themselves intensely political, and set the agenda for policy decisions. Since 1996, for example, the FIFRA SAP has discussed insect resistance management for Bt crops, the Starlink controversy and its implications for food safety (in collaboration with the FDA), the general impact of Bt crops on the environment including genetic transfers and impacts on non-target insects and, most recently, the proposal to register a new type of Bt corn aimed at dealing with root worm problems. Why have these problems been selected? It was not out of scientific curiosity, but because various interest groups have specific concerns, and the scientific forum is an effective way of legitimising certain value choices and in discussing result of experiments and trials which measure how these value choices are being affected.

What is significant, by way of comparison with the discussions covered in the previous chapter on UK's GM scientific advisory committee, ACRE, are those issues that have not been raised at the SAPs. There has been no discussion of the impact of herbicide regimes on food for birds or on the growth of herbicide tolerance – 'superweeds' – topics that have been exhaustively discussed on the UK's ACRE. So it is not a question of science being the basis of policy making. Rather it is about what science is needed to validate, and solve the puzzles thrown up by, dominant value structures which constitute the discourse-paradigm. Even when the scientific evaluation systems in the EU and the USA do focus on similar problems, they are liable to come to different conclusions. I shall discuss, later, how the FDA has effectively advised that use of most antibiotic resistant marker genes is not a problem. The FDA delivered this advice through the medium of a group of experts which it commissioned to study the issue. Yet scientific experts on the UK's advisory committees (and other EU member states) came to rather different conclusions.

What is clearly the case is that no SAPs connected with the EPA, FDA or USDA have demanded that moratoriums on activities be agreed until research has been conducted, as has been the case in the UK and the EU. Moreover the SAPs have exhibited general agreement with the basis of US regulation of GM crops and food that they are not to be treated as being fundamentally different from commercial food and crops. The members of the SAPs appear to be nominated through a political process, that is selected by an agency, but given the procedures involved it does seem that the panels broadly represent the opinions of the scientific disciplines from which they are drawn.

Moreover, statements on GM food and crop policy made by the National Academy of Sciences (NAS) do not take exception to the broad outlines of US GM food crop regulatory policy, and certainly seem to confirm the emphasis of the EPA and other agencies on the matters which are in need of attention. The NAS broadly supports the US approach of not treating GM crops as fundamentally different to conventional crops saying that 'the transgenic process presents no new categories of risk compared to conventional methods of crop improvement' (National Academy of Sciences 2002b: 5) and, with one or two refinements, gives broad support to the USDA's notification system for evaluating safety of GM crops (National Academy of Sciences 2002b: 9). This again contrasts with the EU's system of full risk assessment for all proposals. Similar general support for the EPA's approach to regulation of Bt crops can be seen in another NAS Report (National Academy of Sciences 2002a). I now want to move into greater detail to identify some key statements illustrating aspects of the US GM food and crops regulatory discourse that can be compared with regulatory discourses dominant in the UK and EU system(s). I want to divide up my examination of key discourses in two sections; one before 1998 and another beginning in 1998. As well as analysing the discourses, I want to put these discourses in the context of some key political developments.

Regulatory discourses

All clear for GM food and crops

First I want to talk about discourses on GM food (as opposed to GM crops). The Coordinated Framework implicitly assumes that GM food normally falls under the key category 'Generally Regarded As Safe' (GRAS). This is crucial since under Section 409 of the Food, Drug and Cosmetic Act (FDCA) only food additives will be subjected to regulation and, as the Coordinated Framework says 'If the substance is generally recognised as safe (GRAS) for a given food use, the product is not a food additive' GM food would, normally, only lose its GRAS status if 'the substance has been altered in such a way that it can no longer be recognized by qualified experts to be safe' (OSTP 1986: 21).

Later on this theme was further elaborated with statements such as (referring to cases including genetic modification): 'In such cases, the manufacturer must determine whether the resulting ingredient still falls within the scope of any existing food additive regulation applicable to the original ingredient or whether the ingredient is exempt from regulation as a food additive because it is GRAS' (Food and Drug Administration 1992).

Hence it was left up to manufacturers to decide whether their food was safe, and a voluntary procedure was established whereby manufacturers sent in information to the FDA to justify their product as being substantially equivalent to conventional food, and therefore GRAS. Although not mandatory, by the beginning of 2003, all 53 marketed GM food products had gone through this process. It is a permissive process which rests on an assumption that GM food is safe, and that it does not need to be proved to be so through methods such as animal tests. At the time of the issue of the FDA's original guidelines, in 1992, it seemed that the public was not, by and large, sufficiently concerned to require too much reassurance of its safety as a foodstuff.

The 1992 FDA guidelines on GM food curtly rejected any notion of labelling GM food on the basis that it was, in effect, identical to conventional food. The FDA based this anti-labelling stance on the following logic:

> FDA believes that the new techniques are extensions at the molecular level of traditional methods and will be used to achieve the same goals as pursued with traditional plant breeding. The agency is not aware of any information showing that foods derived by these new methods differ from other foods in any meaningful or uniform way, or that, as a class, foods developed by the new techniques present any different or greater safety concern than foods developed by traditional plant breeding.
>
> (FDA 1992: 13)

This discourse is directly contradictory to the EU's interpretation which treats GM crops and food as being sufficiently different to require labelling. The US courts went along with the FDA's logic. In 1996 the Supreme Court struck down an attempt by the State of Vermont to mandate the labelling of milk derived from cows treated with the growth hormone Bovine Somatotrophin (BST – produced through genetic engineering) (Krenzler and MacGregor 2000: 301).

The question of antibiotic resistant marker genes emerged as an issue during the first big GM crop and food registration debate in 1992 when Calgene formally requested permission from the USDA that their long life GM tomato be given commercial approval, that is, have its status as a 'plant pest' removed. The Flavr Savr tomato, as it was marketed, contained a gene for antibiotic resistance in the form of kanamycin, an antibiotic used, albeit relatively rarely, in clinical practice. In 1993 Calgene petitioned the FDA for specific permission to use kanamycin this way. This use of kanamycin as a marker gene was approved. There was a big debate on the FDA's Food Advisory Committee about kanamycin, but those objecting to the approval of its use in Flavr Savr were heavily outvoted (Nestle 2003: 178).

As Nestle puts it:

> Calgene contended that the kanamycin inactivating enzyme would be destroyed by cooking or normal digestive processes and was unlikely

to function in the intestine. But could the gene for antibiotic resistance jump from food or soil to bacteria in the intestines of animals to people?

(Nestle 2003: 178)

The FDA considered the general issue of kanamycin resistant genes in general, and in a rule issued in May 1994, gave its answer to Nestle's question:

> The probability of transfer of the kan sup r gene (the kanamycin resistant gene) gene to gut microflora is remote and that even under worst case conditions, the number of microorganisms that would be converted to kanamycin resistance is negligible when compared to the reported prevalence of gut microflora that are already resistant to kanamycin ... The agency has determined that exposure to foods that contain the kan sup r gene will not compromise the efficacy of antibiotic treatment because the likelihood of increasing the number of antibiotic resistant microorganisms is extremely low.

(FDA 1994: 11)

In fact this discourse served as the template for judgements about other cases; and approval of use of antibiotic resistant marker genes was routine thereafter. Indeed the CIBA/GEIGY GM soya product using the ampicillin resistant marker gene which was later to cause so much uproar in the EU was approved without any scientific advice being given on the issue of its use of antibiotic resistant marker gene. Attention was more focused on the Bt gene. Indeed the Scientific Advisory Panel of the EPA gave general approval to the use of antibiotic resistant marker genes in 1995 (Krimsky and Wrubel 1996: 151). This is a most crucial example of how scientific uncertainty has been interpreted differently in different places (i.e. the USA compared to most EU member states). These rival interpretations may have been expressed in scientific language but they have had clearly divergent sources of political inspiration and quite dramatically different political consequences.

Was this (US) judgement permitting the use of antibiotic resistant marker genes the result of biotechnological control of the regulatory system? In many ways the biotechnologists DID control the regulatory process, but this only appears to be the case since the discourse that they favoured coincided with a lack of general public concern about the matter. This illustrates the Foucauldian interpretation of power. It is not something wielded by specific actors, but more a system of relations, which in this case delivered results that were to the liking of biotechnologists. It was not because the biotechnology companies were, in conspiratorial manner, suppressing popular feeling. There were simply very few objections made to what was being approved. Only 43 comments were sent in to the FDA about Calgene's petition to obtain permission to use kanamycin as a marker gene, and the large majority of these comments were favourable towards the petition. Indeed there were even fewer comments when the product was proposed for registration with the USDA for commercial cultivation (Martineau 2001: 124 and 147).

After the Calgene case the whole exercise of GM crop regulation took on the appearance of something quite routine. Over 900 permits for small scale field trials had been issued by the USDA by the beginning of 2000. However, over 5,800 field trials had been agreed through the 'notification' procedure. Under this procedure plant breeders simply 'notify' the USDA of the field trial, although the USDA retains the right to demand a full permit application. The notification procedure means that a full assessment is not necessary, usually because the proposal falls into a certain category which has already been given general approval in the context of a previous permit application. (Abramson and Carrato 2001: 248–249).

Unlike the UK's ACRE, the EPA and the USDA have hardly dicussed environmentalist concerns about the alleged dangers from 'superweeds' which have acquired multi-resistance to difference types of herbicide. There has been no insistence that field testing of herbicide tolerant crops be linked to monitoring problems from herbicide tolerance as has been requested by ACRE in the UK. Disucussion has mainly been limited to the 'outsider' groups like the Union of Concerned Scientists (UCS) (Rissler and Mellon 1996). In regulatory circles environmental impact was discussed in terms of the extent to which such herbicide use was reduced, and USDA reports have suggested that herbicide and insecticide use has been generally reduced though use of agricultural biotechnology. On the other hand, there has been some discussion of cross pollination of Bt characteristics with wild relatives of corn, cotton and potato, and also of insect resistance problems. I shall talk about the handling of these issues in the next section on post-1998 discourses.

In 1994 EPA proposals were published for regulating plants that are modified to be virus resistant and also crops modified to produce Bt pesticides. The main issue in the implementation of the draft rules came to be that of insect resistance. Controversy emerged in 1995 in respect of a Monsanto petition to de-regulate a Bt potato. Environmentalists, spearheaded by organic farmers, complained that the innovation could lead to the loss of usefulness of microbial Bt sprays (used by organic farmers) because of increased spread of insect resistance to the Bt. 'Consequently the EPA received several hundred letters' complaining about this (Levidow and Carr 2000: 26).

Levidow and Carr (2000) discuss how 'legal and practical issues' became considerations, and a new practice was instituted to deal with these concerns. It was called 'insect resistance management' (IRM) and involved farmers leaving 'refugias', that is large portions of fields where conventional crops were planted, as a means of slowing down the development of any insect resistance. Companies are required to induce farmers to comply with IRM activities as a condition of the Bt corn being regis-tered by the EPA. The insistence on IRM regimes can be associated not only with environmentalist concern, but also with the perceived interests of the biotech companies themselves who had a natural interest in seeing the commercial life of their products extended by guarding against the development of insect resistance to Bt crops (Abramson interview – 2003). A comprehensive programme of monitoring possibilities for development of insect resistance and also the degree of farmer compliance with IRM programmes has been organised partly by an industry based body called the *Product Stewardship Council.* Nevertheless, the EPA disavowed any notion that Bt resistant crops pose a significant problem. A report to an October 2000 EPA Scientific Advisory Committee commented that:

> Available data indicate that after five years of commercialisation, no reported insect resistance has occurred to the Bt toxins expressed either in Bt potato, Bt corn or Bt cotton products ... EPA believes that significant benefits accrue to growers, the public, and the environment from the availability and use of certain Bt plant-pesticides.
>
> (FIFRA Scientific Advisory Panel 2000a: 13)

Of course, this statement covered not only the issue of insect resistance management but also the thorny topics of threats to native corn species and also threats to non-target insects such as the monarch butterfly. These were issues which arose in the more recent phase of regulation which has involved criticism of GM food and crops.

GM criticism grows

The period from 1998 onwards saw a growth in criticism of GM food and crops in the USA. Although it has not resulted in restrictions on crops or food of the type given authorisation by 1998, there has been an increase in some types of justification required for them, and there is increased sus-picion of new GM food developments. I want to look at some of the (subtle) changes in dominant regulatory discourse that are involved. However, I also want to look at how this happened, by charting events that were associated with the proposals for regulatory change. Some pinpoint events in Europe as being influential: 'In 1999, largely as a response to developments in Europe, public awareness of and opposition to geneti-cally modified seeds and crops did emerge in the United States' (Vogel 2001: 5).

One piece of unfinished business was an FDA re-examination of the antibiotic resistant marker gene issue, an examination that had certainly been triggered, in November 1996, by the furore in Europe over the

CIBA/GEIGY Bt corn. A draft guidance issued by the FDA in 1998 said that 'Marker genes that encode resistance to ... certain antibiotics that are the only drug available to treat certain clinical conditions ... should not be used in transgenic plants' (FDA 1998). Crucially, it did not mention antibiotics that were not 'unique'. This statement effectively gave the all clear to use of antibiotics like kanamycin, and, it seems, ampicillin. Indeed, the guidance has not been transformed into a rule, leaving the decision over whether to follow this piece of advice to the discretion of the biotechnologists. Critics such as Greenpeace argue that there has been no discernable change in practice (Margulis interview – 2003). This guidance on antibiotic resistant marker genes set the tone for other regulatory pieces of fine tuning that emerged in response to the more critical post-1998 climate. They did not actually change the practice relating to GM uses. Essentially they increased the need for more justification and detail in GM products.

We cannot tell how influential Europe was or was not in generating concern about GM agriculture in the USA, although certainly by 1998 it was obvious that Euro-scepticism of GM crop technology ran deep. However, sociological indicators suggest that a domestic dispute over the definition of organic food may have played a part. The 'sociological indicators' to which I refer were increases in activity by campaigners for organic agriculture.

Although the Organic Food Production Act was passed in 1990 it took another eight years before the USDA produced a proposal to standardise a definition of organic food, and when it did, controversy erupted. This was because the proposed rule would have allowed both GM food and irradiated food products to be classed as organic if other criteria were met. Such ideas were anathema to the organic farming and food lobby. An energetic campaign was organised by organic supporters to influence the debate which the USDA staged in early 1998. Indeed by the end of April 1998 over 275,000 comments were sent to the USDA demanding that organic food be GM and irradiation free (Nestle 2003: 233).

What this campaign did do was to build up and energise an army of potential campaigners on what was seen as a closely related subject, that of GM food and crops. Indeed in May 1998 the Center for Food Safety lodged a suit against the FDA for its failure to have safety testing for GM food. This suit was dismissed on the grounds that the FDA did not have a rule that could be legally challenged, only a policy. Nevertheless, as public demands for FDA action on labelling grew, the FDA announced, in October 1999, a series of public meetings to discuss policies on the labelling and also safety assessment of GM food. The FDA received around 50,000 comments on the former issue and 35,000 comments on the latter issue.

There had been criticism of the voluntary notification procedure on new GM food products. A report issued by the Center for Science in the

Public Interest alleged that there were many omissions in submissions to the FDA covering new food products. It was claimed that details of various potential toxicants were not covered in the submissions, that half of the requests by the FDA for additional information were refused and that various data sets had inadequate detail on a range of issues (Gurian-Sherman 2002).

In fact, in a rare instance where there was a distinct difference between a recommendation from a Scientific Advisory Panel and an Administration policy decision, the balance of opinion at an SAP meeting in 1994 come down in favour of mandatory pre-notification (Joint Meeting of the Food Advisory Committee and the Veterinary Medicine Advisory Committee 1994: 63–112). Finally, after the public pressure, the FDA came out in favour of mandatory pre-market notification for new GM food products. It should be noted, however, that this falls well short of the sort of pre-market testing (involving the completion of various safety tests) that GM critics demanded. Indeed it is key point to remember that the concession of pre-market notification still preserves the key notion that GM food products are not fundamentally different to their conventional counterparts, or equivalents. The FDA said:

> The Agency reiterates its view, as stated in the 1992 policy that transferred genetic material can be presumed to be GRAS (Generally Regarded As Safe) ... However, FDA recognizes that because breeders utilizing rDNA technology can introduce genetic material from a much wider range than previously thought possible ... In such circumstances, the new substances may not be GRAS and may require regulation as food additives.
>
> (FDA 2001a)

But had conditions really changed, or was it merely a case that the FDA found itself needing to bend to the wind of increased questioning of the automatic clearance that was afforded new GM food products? Here again we see how political judgements inevitably end up being made in regulatory disputes because scientific uncertainty can be interpreted in differing directions. Political trends had swayed further in the direction of GM scepticism since 1994. However the dominant regulatory discourse was still that the first generation of GM food products were no different from conventional food products and therefore required no special testing regime. It was just that the FDA now wanted information to determine that the product did not constitute an additive that required regulation (and thus safety tests). Nevertheless, at the time of writing, the new proposals on pre-market notification remained as a draft rule only.

The second thorny issue which came up in the FDA's 1999 hearings was labelling. Most of the written comments that the FDA received wanted mandatory labelling of GM food, or in FDA-speak they 'requested mandatory disclosure of the fact that the food or its ingredients was bio-engineered or was produced from bioengineered food' (FDA 2001b). Once again, the FDA stood its ground in its attitude that:

> Many of the comments expressed concern about the long term consequences from consuming bioengineered foods, but they did not contend that any of the bioengineered foods already on the market have adverse health effects. The comments were mainly expressions of concern about the unknown. The agency is still not aware of any data or other information that would form a basis for concluding that the fact that a food or its ingredients was produced using bioengineering is a material fact that must be disclosed under sections 403(a) and 201 of the (Federal Food, Drug and Cosmetic) act. FDA is therefore reaffirming its decision to not require special labeling of all bioengineered foods.
>
> (FDA 2001b)

The FDA did decide to allow voluntary labelling to say whether food is produced using, or not using, GM ingredients. However there was a contrast between the labelling that was allowable for the purposes of promoting GM foods and the labelling that was allowable for the purposes of promoting non-GM food. Labelling can promote GM food by saying, for example, that high oleic acid soybean oil from GM soybean will decrease saturated fat content (hence implying that it is superior). On the other hand producers opting to say that their food was not produced using GM ingredients cannot imply that their product is superior to GM food.

The issue of whether GM food is different to non-GM food is a hotly contested issue, yet it is a fundamental truth of FDA's GM food discourse that it is not different to conventional foodstuffs. Thus, according to this truth system, not only is GM food intrinsically safe but attempts to imply anything else are illegal. The truths expressed in this regulatory discourse coincide with those of the biotechnology and farming interests rather than the GM-sceptical interests, and even (as I shall discuss later) rather than those of the food processors and manufacturers.

Buoyed up by opinion polls which appeared to show that a large majority of voters wanted mandatory labelling of GM food, anti-GM food activists began to campaign for state-based ballots on the subject. In Oregon activists managed to get a question onto the November 2002 election ballots and they were optimistic for a positive endorsement of mandatory labelling. However, pro-biotechnology interests argued forcefully that labelling would put up farm-produce prices by 30 per cent and the Oregon poll recorded a 70:30 verdict against the idea of mandatory labelling.

Pro-labelling campaigners blamed the result on the large disparity in the funds available to their campaign compared to the resources spent on

TV adverts opposing the labelling case. They said this meant they were unable to effectively contest what they saw as exaggerated claims about price increases that would follow segregation of GM food and other food. Nevertheless, the fact remains that the voters kicked the labelling campaign into touch by a large majority. The fact that they did underscores the results of the more rigorous comparisons of EU and US food on GM food and crops which indicate greater European scepticism of GM food technology compared to the USA.

Certainly, Oregon voters seemed more concerned about claims of food price increases compared to rival claims of uncertainty about food safety and impacts on the environment. In this case a discourse of food price increases has overcome a discourse focusing on safety and environmental concerns. We can see the influence of economic structures in that biotechnology companies and many farmers, who between them have a lot of money tied into GM crop products, saw their interests as being upheld by putting considerable effort into defeating proposals to label GM food products. They feared that this may lead to boycotts of GM food. However, even these economic structures are not as material as they seem (as opposed to being ideas in discourses), for some farming interests, such as the American Corn Growers Association (ACGA), support an alternative perception of farmers' interests by urging farmers to grow conventional food crops. The ACGA supports labelling of GM food. In reaction to claims that labelling GM food will produce 30 per cent increases in farm prices they argue that:

> I think that they've inflated the cost because they don't want to do it. The United States Department of Agriculture continues to come out with these outrageously high numbers for implementation of the programme. We already have organic labelling in place. I go to several grocery stores where the label tells me not only that it is organic but where it came from. Their prices aren't any higher.
>
> (Mitchell interview – 2003)

As I discuss later, the failure to segregate GM and non-GM crops mean that US corn and soya farmers are effectively barred from EU markets since their produce would have to be labelled as 'GM' even if it was not produced that way at the farm. This is one reason why the National Corn Growers Association has urged the US government to make a complaint to the WTO about the EU's slowness in licensing various GM crops. Others might argue that it would serve US farmers better to segregate the crops and at least have the possibility of sales in the EU.

The debates about GM food and crops in the UK and the USA have both involved arguments about the impact on wildlife, although the take on these wildlife impacts has been different. Again, we can trace the different interpretations of wildlife impact to path-dependent factors. In the

UK the centre of concern has been the impact on food availability for birds, something which parallels the strength of the bird protection movement in the UK. In the USA, concern has instead focused on insects. I refer the reader to the quote from the final paragraph that Carson wrote for her seminal text, *Silent Spring* (Carson 1962: 257). Carson spoke about the human industrial monster turning against the insects, and it was her campaign against pesticides that kick-started what is regarded as the modern environmental movement in the USA. It is hardly surprising, therefore, that as criticism of GM crops emerged in the USA in 1999, researchers thought about testing the impact of Bt crops on insects, in this case the much loved monarch butterfly.

Soon after the publication of the studies suggesting that Bt pollen harmed monarch butterflies the EPA went about seeking studies of impacts on non-target insects and it issued a precautionary notice requesting that non Bt corn refugias should be placed between Bt corn habitats and the Bt corn. Nevertheless the EPA concluded that:

> In light of recent environmental effects concerns from commercialisation of Bt crops the Agency has reviewed new and existing data regarding non target wildlife effects for Bt crops with a special emphasis on Lepidoptera and monarch butterflies ... The weight of evidence from the reviewed data indicate that that there is no hazard to wildlife from the continued registration of Bt crops.
>
> (FIFRA 2000b: IIC69)

Later EPA studies confirmed these conclusions, although critics such as Jane Rissler from the Union of Concerned Scientists remained unconvinced by the tests. As one of the Panel members commented, though, the tests on impact on target insects were made in comparison with the sort of microbial Bt sprays which are used by conventional (and organic) farmers rather than with crops that are not subjected to Bt toxins at all (FIFRA Scientific Advisory Panel 2002). The point that can be made here by comparison with the UK experience is that there was no moratorium imposed on Bt crop planting while tests were carried out.

There was less closure, from a policy point of view, regarding the issue of 'outcrossing and weediness' of Bt crops, or as the European anti-GM activists put it 'genetic pollution'. As I discussed in Chapter 1, this has emerged as a major controversy with regard to allegations that Bt corn was crossing with wild Mexican corn (from which commercial versions had originally been developed). The EPA itself was concerned about the possibility that Bt cotton may cross with wild versions of cotton Florida and Hawaii to the extent that the EPA actually restricted sales of Bt cotton in these areas (FIFRA 2000b: IIC1) The subject of gene flow from Bt plants is an area which has been identified by the National Academy of Sciences as an important topic for further research, and an area where the USDA and

EPA could improve their co-ordination of policy (National Academy of Sciences 2000). However, to date, there appears to be less concern about the fate of Mexican corn, which is, admittedly (for the EPA) beyond its jurisdiction.

By the end of 2002, however, what may amount to a far more explosive debate (in the USA) about the future of agricultural technology emerged with the publication of a joint FDA/USDA Draft Guidance for Industry (Food and Drug Administration/US Department of Agriculture 2002) on growing crops for the purpose of producing plant-based pharmaceutical products (bio-pharming). It did not augur well for the biotech companies interests that the publication of the draft guidance coincided with the ProdiGene affair which I discussed earlier.

A range of interest groups including the influential Grocery Manufacturers of America (GMA) have expressed concern about the possibility of contaminating food supplies. The Draft Guidance itself, while discussing a range of precautionary measures that the industry should take, stopped short of acceding to their demands for confinement to greenhouses and avoidance of use of food crops (Grocery Manufacturers of America Comments 2003). The nearest it came to these demands was to say:

> When a plant species that is used for food or feed is bioenginered to produce a regulated product, you should consider the use of strategies that allow the bioengineered pharmaceutical plant line to be readily distinguished from its food or feed counterpart . . . You should also consider strategies to reduce the likelihood of unintended exposure to a regulated product by restricting the expression of the bioengineered pharmaceutical product to a few specific plant tissues . . . **For such plants that outcross, you may want to grow them in regions of the country where little or none of its food/feed counterparts are grown.**
>
> (Food and Drug Administration/US Department of Agriculture 2002: 9)

I have added the emphasis on what many would see as the most significant section of this piece of advice about avoiding certain parts of the country, but this advisory statement goes a long way short of what the food manufacturers and distributors want. I shall comment more on this topic in the next section on policy networks, but it is worth saying here that the biotechnology industry is likely to strongly resist calls to use non-food crops, since, for example, corn is around three times cheaper as a base for pharm-crops compared to tobacco (Pew Initiative on Food and Biotechnology 2002: 5). Confinement in greenhouses would also add a considerable layer of costs. However, following the ProdiGene affair, the Bio Industry Organisation did announce a voluntary moratorium on growing pharm-crops in the Corn Belt.

The first petition to de-regulate (that is, commercialise) a pharm-crop is yet to be heard, and the controversy surrounding this type of biotechnological application is likely to be bigger than those which have previously been de-regulated. A further area of controversy is GM animals, a topic which is being treated with caution by the FDA. Indeed the FDA has decided to regulate GM animals as an animal drug, and thus the regulatory procedure is lengthy. The sceptics argue that this is unsatisfactory since the drug testing regimes shields information behind a veil of commercial secrecy. A type of salmon modified to grow faster (and thus bigger), is the first GM animal to be processed. Aquabounty, the company who have produced this innovation, do not expect to hit the market place before 2005, and other GM animals are said to be a decade away. Even after FDA approval there will be pressure for the GM salmon to be considered by the National Fish and Wildlife service who will consider complaints by environmentalists that the salmon could endanger existing wild salmon species. The GM salmon are reared to be sterile, but critics say that this may not be foolproof.

Other new GM plant developments are equally controversial. A proposal by Monsanto, to commercialise a grass that is tolerant to Ready Roundup herbicide, has run into a counter petition by the Center For Food Safety who argue that the product could end up being a weed. The Center is also contesting the holding of trials for GM insects that will encourage sterility in crop pests.

The trade dispute discourse

One USDA discourse that has been a recurrent theme since 1998 is that the EU has deployed restrictive trade measures to block imports of GM crops and food products from the USA. Finally, after many threats, the USA made a formal complaint to the WTO in May 2003 about the EU's delay in approving GM food and crop products. The complaint was made in collaboration with seven Latin American countries plus Australia and New Zealand. The USDA said that:

> The EU's persistent resistance to abiding by its WTO obligations has perpetuated a trade barrier unwarranted by the EC's own scientific analysis, which impedes the global use of a technology that could be of great benefit to farmers and consumers around the world ... (The EU's GM crop and food) approval moratorium is causing a growing portion of US agricultural exports to be excluded from EU markets and unfairly casting concerns about biotech products around the world, particularly in developing countries.
>
> (USDA News Release 2003b)

US agricultural exports to the EU have certainly been badly hit by the GM food dispute. US corn exports to the EU, worth about one per cent of total US corn production in 1995, fell to practically zero after 1996. US soya imports do not have the problem of being mixed with unauthorised GM products (unlike US corn produce), since there is only one type of GM soya in use and that has been authorised by the EU. However there has still been increasing pressure from EU food retailers to source soya from non-GM soya crops.

The American farming and farm distribution industry, as part of their opposition to mandatory labelling, refuse to segregate GM and non-GM produce. This means that even non-GM soya farmers cannot sell their produce to the EU if the EU customers are insistent on their wish to buy only non-GM soya. If there is no system of labelling US GM soya, then the EU buyers will not place orders for American soya since they do not know whether they are buying non-GM soya; most probably they would be buying a high proportion of GM soya from the USA since a range of soya sources are mixed together in shipments from the USA. Indeed, EU imports of US soya have been falling because of general opposition to GM foodstuffs. In 1995 around 10 million tons of soya was exported to the EU, but this had fallen to 6.3 million tons by 2000 (Van der haegen 2002). Total US soya production is around 58 million tons, so the loss of the EU export market is a major worry for the US soya industry. Since the retail boycott of GM food became effective in 1998, GM food exports to the EU have mostly been restricted to the animal feed markets. As I shall discuss in the next chapter, even these exports are under threat because of new, tighter, EU labelling regulations. At this point rational choice theorists might step into the picture.

Rational choice theories use game theory to predict outcomes based on modelling how actors might pursue their self-interests. However, this depends on a variety of factors that may be difficult to judge in advance, such as (in this case) the reactions of actors in other countries. Do American farmers rely on expectations that increasing prevalence of GM crops will ultimately force EU food buyers to accept GM supplied because of the lack of any alternative? In fact, as I shall point out later, this does not appear to be happening. Farmers in various parts of the world appear are becoming resistant to growing GM crops precisely because this will block them out of EU markets. China's earlier ardour for GM crops has cooled considerably and it has become reluctant to sanction them because of access to EU markets (Checkbiotech 2001). Indeed, because non-GM supplies of soya now demand a premium price (because they are required by EU and other anti-GM markets) farmers now have a positive incentive to produce non-GM rather than GM crops, thus negating any cost-cutting advantage of growing GM crops in the first place. China has even declared North-Eastern areas as a GM soya free zone. Alternative (to US), non-GM sources of soya for the EU have generally come from Brazil, but China and India are also potential exporters of soya to the EU. As I shall discuss in Chapter 5, farmers in countries awaiting EU entry are not keen to see GM crops threatening their supplies of non-GM canola to the EU.

Few biotech lobbyists seem to believe that the WTO complaint will reverse this trend of falling EU farm exports. Many US farming leaders are alarmed by the possibility (in fact reality) that the EU's reluctance to sanction GM crops and food products, and demands for strict labelling of GM food products, is spreading to other countries, thus hitting US agricultural trade further. The WTO complaint, therefore may be classed as a measure 'pour encourager les autres', that is to dissuade other countries (including China) from taking the EU's GM-sceptical route. The problem for the US complaint is that it is too narrow. To be effective it has to focus on labelling and issues such as regulations governing crop separation distances between GM and non-GM crops, which can also serve to restrict cultivation of GM crops. Yet, as I shall explain in Chapter 5 a complaint about labelling is not likely to be successful.

Rational choice theory may offer a framework upon which some stories may be hung in order to help explanation (Toke 2002b). It may be the case that some perceptive analysis could be done in this case using restricted notions of how actors may interpret their self interest. However rational choice theory is limited where (as is usual in politics) differing interpretations of self interest are possible and actors have more than one choice which can be used to pursue their self interest. For example, US farmers have to choose whether it does or does not serve their interests to segregate their crops into GM and non-GM. Second, there are difficulties in predicting how other actors (including for example, Chinese farmers) will interpret what is in *their* best interests. Much rational choice theory fails to live up to its own high standards of generating universal laws and instead produces post-hoc rationalisations (Green and Shapiro 1994). Discourse analysis makes no predictive claims, but at least it can attempt to examine the environment in which self-interests and events have been constructed and interpreted.

A common discourse in the USA is that the motivation (as well as effect) of the EU's policies on GM food is protectionist. For example, one commentator says:

> It's clear that the EU ban is not a safety precaution, but a barrier to trade. The EU is citing phony safety concerns to protect its farmers from competition and to protect its farmers from competition and to protect its system of bloated farm subsidies.
>
> (Bailey 2002)

The problem with this sort of attitude is that, by failing to understand the roots of EU policies, the US agricultural sector may be treading a path that will severely damage its own interests. US farmers may be waging what they see as a war against protectionism through appeals to the WTO and, perhaps later, through trade sanctions. Yet, this strategy is unlikely to reap success so long as the EU and other countries insist on mandatory labelling of GM food products. US farmers may well protect their trade interests better by 'going with the flow' and agreeing to segregate their produce at source, although Canadian farmers are suggesting that, in the case of oil seed rape, even this is difficult because GM and non-GM seeds tend to become mixed up together.

Policy networks and GM food and crops

The US system, with its rather more decentralised government (both in terms of its Federalism and separation of powers), is not as good a candidate for the use of the type of Marsh/Rhodes (1992) policy network analysis that I applied to the UK in Chapter 3. As suggested later, the concept of an 'iron triangle' may be more appropriate in some respects of the GM food and crops case. Nevertheless, as Nestle points out, farmers and food producers have, since World War Two, constituted a powerful policy community (she calls it 'agricultural establishment') in terms of deciding agricultural policy. This influence has declined since the 1970s as new constituencies have appeared to lobby Congress (Nestle 2003: 97–99). It is possible to identify the main networks involved in the GM crop and food policy area, and certainly identify those groups that have been most influential.

The Food Chain network

The farmers, seed and grain distributors, biotechnology companies, agrochemical companies, food manufacturers, retailers and food processors all meet together, on a weekly basis in a so-called 'Food Chain' policy group. When in agreement, this network has been virtually unstoppable in translating its requirements into regulatory action. The environmental groups who are sceptical of, or who campaign against GM crops and food are not privy to this policy network.

There is a broad consensus among the 'Food Chain' that GM food should not be regarded as any different to conventional food and that there should not be mandatory labelling of GM food. Moreover there is agreement that EU policies on GM food are unscientific and protectionist. On food labelling Val Giddings of the Biotechnology Industry commented that 'It is a fraud on the public to boycott products using a label implying that the food is safe by saying that it is non-GM' (Giddings interview – 2003). On EU policies Karil Kochenderfer of the Grocery Manufacturers of America said: 'The labelling and traceability scheme that is proposed in the EU is non-feasible, impractical and fundamentally flawed' (Kochenderfer interview – 2003). The Food Chain urged the US government to resist EU efforts to codify the 'precautionary principle' in WTO regulations during the 2001 Doha round of trace negotiations (Grocery Manufacturers of America Correspondence 2001).

It is, perhaps, appropriate to describe the influence of the 'Food Chain' network as sometimes being akin to the sort of influence conjured up by the notion of an 'iron triangle' (Cater 1964). In this notion a particular set of interests captures the decision making process through its links with two other concerns of a policy triangle, namely members of Congressional Committees and Department officials. Hence all other interests are frozen out of the policy process.

However, within the 'Food Chain' there are differences on some policies. Hence it is useful to divide up the Food Chain into two categories – on the one hand there are the biotechnologists and the farmers and on the other hand there are the food processors and manufacturers.

Biotechs and farmers

I have already described some leading biotechnology companies in Chapter 3 such as Monsanto, Syngenta and Bayer-Aventis. They are represented by the Biotechnology Industry Organisation, and they also tend to overlap with the agrochemical companies who market what they call plant protection products and what many other people call herbicides and insecticides. The approach of these companies has (as in the case of the UK) been similar to that of the main farming organisations such as the National Corn Growers Association (NCGA), the American Soya Association and the biggest of them all, the American Farm Bureau. The NCGA, and other farming interests (and their supporters in Congress) have been loudly supportive of the US government's decision to make the formal complaint to the WTO about EU GM food and crop policies. The biotech companies and the leading farm organisations are the most enthusiastic proponents of the pro-biotechnology storyline that GM crop technology provides food that is not only at least as safe as conventional food but that also it is among the latest in a long line of technological developments that promise to make food healthier and reduce the environmental impacts of food production.

The approach of the biotech and farming interests deviates from the food manufacturers and processors on two key issues. First is labelling. While the biotech companies are steadfastly opposed to mandatory labelling they support (in contrast to the grocery manufacturers) what they call voluntary 'informative' labelling. For example the biotechnology companies want the freedom to label what they see as 'healthy' aspects of GM crops such as a high oleic oil GM soya product that is being marketed.

The second issue which divides the 'Food Chain' policy network is that of plant-based pharmaceuticals, or pharm-crops. 'It's difficult to draw the line between pharmaceutical products and enhanced healthy products' claims Val Giddings of the Bio Industry Organisation. He continues:

'There are a host of things that can be done to avoid problems such as occurred with ProdiGene and the pig vaccine. The FDA can give approval to certain plant based pharmaceutical products to show that the adulterated crops will do people no harm' (Giddings interview – 2003). USDA and FDA policy seems to have closely followed the contours of the interests of the farmers and biotech companies. However, there are signs that the Bio Industry Organisation is bowing to pressure from the food manufacturer's lobby who are themselves sensitive to the possible emergence of widespread consumer concern on the issues of pharm-crops and, perhaps on GM animals.

Food manufacturers and processors

The most publicly prominent organisation in this lobby is the Grocery Manufacturers of America (GMA). In advancing its concerns about plant-based pharmaceuticals it has networked with groups like the American Bakers Association, Biscuit and Cracker Manufacturers Association, the International Dairy Foods Association, the National Confectioners Association, the National Council of Chain Restaurants, the National Restaurant Association, the National Soft Drinks Association and the Institute of Shortening and Edible Oils and the Food Marketing Institute (Grocery Manufacturers of America News Release 2003).

The retailers should also be associated with the views of this lobby. The Food Marketing Institute, which represents retailers and wholesalers, has collaborated with the GMA on biotech issues, but in general the retailers have left the articulation of their interests up to groups like the GMA. This retail quiescence is in marked contrast to the profile of the retail industry in the EU.

The food manufacturers/processors lobby is very powerful and for the major part of the GM crop and food debate it has spoken from the same script as the farmers and biotech companies. They regard GM food as a progressive technical development. However they are cautious about pharm-crops, as discussed earlier. They call for a 'comprehensive' rather than 'piecemeal' regulatory system for pharm-crops (although not conventional GM food crops) (Grocery Manufacturers of America news release 2003). Interestingly, this sounds like the EU's philosophy underpinning GM crops in general! The food manufacturer's verdict on the use of GM animals for food production will depend on consumer responses. The food manufacturers are closer to the food market place than the farmers and biotechnology companies, and the food manufacturers want to ensure that chemical crops or GM animals do not erode consumer credibility in the products that they market.

The GMA also deviates from the biotech and farming interests on labelling in that they are sceptical of voluntary labelling as well as mandatory labelling of GM crops. 'Whether it is pesticides, whether it's soil

erosion, whether it's animal welfare you have a variety of sources that consumers can get information but we reserve the label for the most critical
information, and that is product safety' (Kochenderfer interview – 2003).
However, they appear to have been mainly defeated on this issue in defer-
ence to the demands of both biotechnology companies, and, to a circum-
scribed extent, the demands of GM-sceptical groups for voluntary
labelling of food with respect to genetic modification.

The GMA and other food manufacturers are pragmatic in their general
support for agricultural biotechnology compared to the fixed interests of
the biotechnology companies. They have shown some concerns about
consumer reaction to pharm-crops and they have certainly cut their cloth
to suit their interests when it has come to selling to Europe. Although
sometimes critical of European food distributors and retailers for caving
in to anti-GM pressures they have (after some pressure) co-operated with
European demands for exports to Europe to be sourced from non-GM
crops.

The GM sceptics

The term 'GM sceptic' covers quite a wide range of people and the
alliances between them tends to shift according to the particular campaign or court case that is being pursued. It would be wrong to describe
them all as 'anti-GM', and indeed this 'discourse coalition' could be said
to include much of what passes for majority opinion in the EU. There are
big differences of emphasis among these groups, although there is a clear
common storyline involving calls for more thoroughgoing checks on GM
food safety and environmental impact and also for mandatory labelling of
GM food. Broadly speaking we can divide the sceptics into two groups; the
pragmatists and the anti-GM activists.

The pragmatists include more establishment-oriented groups such as
the Union of Concerned Scientists, the Environmental Defense Fund and
the Consumers' Union. These groups argue that GM crops can sometimes
be justified, but that there needs to be much tougher regulation, including specific legislation covering GM crops and mandatory scientific testing
and labelling of GM food. Their approach has been supported by a
handful of Congressmen led by Representative Dennis Kucinich who has
(since 1999) proposed Congressional Bills to give effect to these demands.
However the Bills have not had sufficient support to be given priority by
the relevant Congressional Committees. The demands of the 'pragmatic'
GM sceptics are strikingly similar to the actual EU regulatory regime covering GM crops. The American Corn Growers Association, which claims to
have about half the number of members compared to the National Corn
Growers Association, has adopted a GM-sceptical approach and supports
causes such as mandatory labelling of GM food.

The anti-GM activists include what might be called the 'usual' green suspects, Greenpeace and Friends of the Earth (FOE) as well as the Foundation for Economic Trends which activised in earlier days. The approach of the radicals has been to try to organise boycotts of GM food, persuade food retailers and manufacturers not to stock or use GM food, and launch legal cases against the Administration. A range of groups have campaigned at a state level for labelling of GM food.

FOE and Greenpeace work with a variety of campaigning groups, mainly organised on a state-by-state basis. FOE's biggest campaigning success was the exposure of the Starlink problem. Greenpeace succeeded, in 1999, in pressing Gerber's to stop using GM food in its baby food, and through the 'True Food' Campaign in pressing the upmarket 'Trade Joe's' chain of stores to stop selling GM food. Greenpeace has supported the work of the Center for Food Safety which has been the main legal arm of the anti-GM movement. The Center has submitted petitions and launched court cases promoting safety testing of GM food, criticising the registration process for Bt corn crops, pressing for tougher regulation of GM animal products and attacking the notion that GM grass and insect products be given commercial authorisation.

Other radical groups include the National Farm Coalition and the Organic Trade Association which represents the small, but growing, organic sector. The 400,000 strong organic Consumers Association has provided a lot of support to anti-GM food campaigns.

Conclusion

We can now summarise some key discourses that underpin the US regulatory system covering GM agriculture. These can be compared with the discourses that underpin EU and UK regulatory approaches

The first, most basic, principle underpinning US GM agricultural regulation is that GM agriculture is not regarded as being fundamentally different to conventional agriculture, but merely represents the latest development in a long line of progressive technical evolution. This underlying philosophy is in keeping with the discourse of the biotechnology industry and has the regulatory consequence that existing laws are used, sometimes creatively, to regulate GM crops and food. It is this underlying philosophy and legal approach that is sharply distinguishable from the general approach of the EU.

A second key discourse underpinning the system is the free market approach, namely that the state should stick to safety concerns and should not be involved in commercial issues. This is associated with confidence in the (biotechnology and farming) industry and also with a third key discourse, that of trust in regulators. Again, this 'trust' discourse is discernably different from the attitude towards regulators in the EU. It is at this point necessary to point out that this conflicts with textbook theory concerning the formal difference between European and US environmental regulation. The US system, developed at the end of the 1960s, does involve a clear separation between risk management and risk assessment and the possibility of the regulators being challenged in court. This is in contrast to the traditional UK approach of 'consensus' between industry and regulators. However, this legal distinction says nothing about the contemporary influence of environmentalists in the USA compared to the EU. Today, in political and regulatory terms, we can say as a broad generalisation (one that is much typified by the GM food and crop case) that environmentalist discourses appear to have wider resonance in the EU compared to the USA.

A fourth key discourse concerns food safety in that because GM foods are no different to conventional foods, the US system involves no special checks on GM food safety, merely a demonstration that the food is 'substantially equivalent' to conventional food (in contrast to the EU, as we shall see). Indeed even this demonstration has been on a voluntary basis. Fifth, mandatory labelling is ruled out for GM food (in contrast to the EU) again, on the basis that GM food is no different to ordinary food. However, voluntary labelling is permissable provided that such labelling is not used to disadvantage the marketing of foods manufactured from GM food.

A sixth key discourse concerns use of antibiotic resistant marker genes. In contrast to the EU, there is no legal barrier to their use in GM crops, and US regulatory guidance leaves final decisions on this issue to the biotechnologists. A seventh discourse concerns the impact of GM crops on herbicide use, which is that herbicide use will be reduced (in the case of GM herbicide regimes) without any significant side-effect. The notion of herbicide tolerant superweeds has been the subject of much discussion in the UK and the EU, but in the USA it has hardly been raised outside the confines of briefings issued by environmental organisations.

However, there has been a discourse about gene flow from Bt crops. The dominant discourse is that this is a concern and requires further research and also regulatory action to limit sowing of some Bt crops in certain areas.

A ninth, wildlife impact discourse, has emerged following the controversy over the monarch butterfly, but the dominant EPA discourse is that Bt crops present no threat to non-target insects. However, the EPA has developed a strategy for avoiding the growth of insect resistance to Bt toxins through the 'refugia' system.

The wildlife impact discourse is path dependent in the sense that it reflects earlier concerns laid down in the 1960s about the impact of pesticides on insects. This wildlife discourse is to be compared to the biodiversity discourse in the UK which has been concerned with the impact of herbicide tolerant GM crop regimes on the food for birds. I can find no trace of US regulators (or indeed, even, of NGOs) having discussed this issue.

Policy on newer developments such as plant-based pharmaceuticals and GM animals is still being formed. Such policies and justifications that have so far emanated from the FDA and USDA on these issues coincide more with the views of the biotechnology companies than with the food manufacturers, but the issue remains in a flux.

The set of dominant discourses that I have discussed reflects the concerns of the dominant policy network, the 'Food Chain' network. The GM sceptics have had very little influence on the policy process, or on scientific judgements about GM food and crops. They have had a small amount of success in lobbying a few companies to avoid GM sources when those companies, such as Gerber's or McDonald's, have become especially sensitive about the 'natural' image of their products. Even within the 'Food Chain' network it seems that the alliance between biotechnology and the majority farming interest have, so far, gained the upper hand in conflicts with the food manufacturers and processors. However, if there is a series of contamination problems with pharm-crops, perceived ecological problems with GM salmon, or consumer sensitivities about GM animals, then the demands of the Grocery Manufacturers of America may be accepted more comprehensively by the regulatory agencies.

That the future of biotechnological development into pharm-crops and GM animals is uncertain is mainly a function of concerns about consumer reactions. However, I must remind readers of the emphasis placed earlier on in the chapter that the apparently greater acceptance of GM food products by consumers compared to the USA is based on a rather different attitude to food held by US consumers compared to the Europeans. I shall say more about this in the next chapter when I discuss European attitudes to food.

The US system purports to separate out science and politics through the division between risk assessment and risk management. No such distinction is possible, however. Some discourses (which underpin regulatory policy) have been argued in relatively technical terms, for example debates about antibiotic resistant marker genes or about impact of Bt crops on non-target insects. Here the scientific advisory committees have been important in 'translating' the values debate into scientific language (and also vice versa) and also in estimating types of risk according to pre-stated criteria. However, it is the 'pre-stated criteria', the value systems, the cultural interpretations of scientific uncertainty, the notions of what issues are important, that are the key to the decisions that are taken. These value systems are reflections of those of the networks that inhabit and dominate the policy arena.

The scientific advisory bodies reflect the value priorities of these policy networks, and the composition of these bodies will also generally reflect the contours of the dominant policy networks. That they do so is not an act of officials who are acting in an unusually conspiratorial manner, but merely a reflection of two trends. First, normal practice of recruiting scientists who will work with the general aims (and therefore policy discourses) of the regulatory agencies, and second the trend towards scientific and technical elites drinking from the same pool of cultural values that infuses the interest groups that comprise the dominant policy networks. The discourses which are dominant among the scientists are closely associated with the discourses which justify and comprise the regulatory arrangements and which make up the discourses of the dominant policy network, the Food Chain. These discourses constitute the USA's discourse-paradigm on GM food and crops.

This does not mean that nobody outside the dominant 'food chain' policy network has had any influence. As was pointed out earlier, Rifkin's court challenges in the 1980s (and associated campaigning) was associated with the establishment of some interventionist regulation of GM crops, even though it was tied to existing laws.

I have already discussed the dominant regulatory discourses concerning GM food and crops in the UK (in Chapter 3), but the key issues of UK GM food and crop politics are, in any case, decided at the EU level. Hence it is now important to move on to look in greater detail at the EU's system of GM food and crop regulation.

5   EU cools on GM food

This chapter needs to be read in conjunction with Chapter 3, on the UK, as a means of comparing GM food and crop politics in Europe and the USA. Chapter 3 was a more in-depth investigation of the GM food and crop politics in one state, although the UK's legal framework for regulating GM food and crops is largely determined at an EU level. Indeed many arguments, such as the arguments over labelling GM food, can be most effectively dealt with at the EU level, which is one of the things I do in this chapter.

As with the two previous chapters, I use discourse analysis and network analysis to examine regulatory politics. However one difference between this chapter and the previous two chapters is that I want to devote a lot of time to discussing the relationships between, and sometimes within, the different EU institutions. This is important both as a means of understanding how GM food and crop policy has been decided in the EU, but also as a means of furthering understanding of the workings of the EU. Theory developed on the EU in Chapter 2 will be used to guide this venture.

I shall begin this chapter, like Chapters 3 and 4, with a review of the politics of food particular to the case study, in this case the EU. Following this I look at the arguments surrounding the notion that the BSE crisis was the principal contributor to distrust of GM food in the EU. I then want to examine the regulatory discourses on GM food and crops in legislation that was passed in the period until 1998 – a temporal division that I have applied to my discussions of regulatory discourses in the previous two chapters. I shall move on to discuss the changes in discourses apparent in legislation approved, and (at the time of writing) about to be approved, and then look at the system of scientific advice in the EU. Finally I shall move on to look at the networks of groups that have been related to the various main storylines on key issues, especially labelling, First, the politics of food.

Food – a matter of history and origin?

In Chapters 3 and 4 I challenged the notion that the differences in food politics (and attitudes to food risks) in the EU and the USA are mainly explained by differences in food 'events' or regulatory competence. A positivist approach will emphasise the food 'crises' in Europe as causes of distrust of regulators and of GM food. The most important food crisis has been the BSE crisis, as described in Chapter 3, as well as other affairs such as the British salmonella-in-eggs affair (also mentioned in Chapter 3) and the Belgian dioxin crisis. The Belgian dioxin crisis arose in the middle of 1999. It involved the feeding of animal fat containing carcinogenic dioxins to farm animals with consequential dioxin contamination of meat products. Two cabinet ministers resigned in the controversy. Also in the summer of 1999, there was a recall of millions of bottles of Coca Cola in Belgium after many schoolchildren complained of stomach pains after drinking the supplies. However, no toxicity was ever established in this case. As I argued in Chapter 4 (and will expand upon later) a 'positivist' view would argue that GM food was suspected because of 'events' such as these. An interpretivist approach would examine how similar events could be interpreted differently at different places and times. In Chapter 4 I argued that there was plenty of possibilities for food 'crises' in the USA, but attitudes to food and regulators are different there.

In Europe, food is a major signifier of identity. As one account puts it: 'In many European societies food is crucially linked to a sense of belonging to a national community and to the ways each nation has customarily portrayed itself, and often, derogatorily, the "others"' (Sassatelli and Scott 2001: 215). The British have a tendency to refer to the French as 'The frogs'. Meanwhile the French have been known to call British people 'les rosbifs', and since the BSE crisis 'la vache folle'.

Certainly, one can hardly imagine the French, the Italians and the Greeks feeling particularly at ease if a food innovation is constructed as something that will undermine the national sense of identity that is embodied in their food and the treasured cuisine that is traditionally cooked at home. These three countries show high rates of public opposition to GM food in Eurobarometer surveys, which I will examine in more detail later on. The Americans, by contrast, see themselves as a melting pot of many cultures and in recent decades have exhibited a marked tendency towards eating out rather than at home.

The anti-GM mood in Italy, France, Greece, and also, to a great extent in other EU countries can be associated with the emergence of groups such as the 'Slow Food' Movement and a general reaction to 'junk food'. For many people, the chief source of this 'junk food' is the USA. The Slow Food Movement is particularly strong in Italy. The Slow Food Movement derives much of its identity (to use a notion developed by Derrida) in terms of what it is **not** (Derrida 1976: 313–316, cited by Howarth 2000:

37). Slow food is not fast food. Slow food is defined in a specific way to draw out a construction of what is good about quality food as opposition to bad production of corporate 'fast food'. As a quote from the 'Slow Food' website puts it:

> Fast food doesn't necessarily mean eating fast, but it does mean having no time to savour what we eat and find out about its history and origin. For us, slow food means learning how to manage our time, savouring food with pleasure and awareness. It is not so much the quantity of time we spend at the table that counts as the quality of the food we eat and the relationship we establish with it.
>
> (Slow Food FAQ 2003)

Note the key references here to 'history' and 'origin'. The relation to GM food is precisely that GM food is a product, born out of the latest technology, that has no history, and moreover has no specific place of origin. By contrast, 'good', or 'slow' food not only has links with the past but also it originates from a specific place or region. 'Slow food' is not necessarily organic food, but one can see how organic food slides easily into the ideas conjured up by slow food because of the organic appeal to tradition and also the ecological concern for the consumption of locally produced food.

José Bové, a small French cheese farmer-producer, has become the French personification of the campaign against fast food and what is seen as US-inspired corporate globalisation. He has articulated the notion of *la malbouffe*:

> This literally means 'bad food' which he (Bové) equates with GMOs, Mc Do, and all the products of globalised culture and agriculture . . . Translated imperfectly into 'bad chow', or 'junk food', la malbouffe symbolizes for Bové everything distasteful about globalisation ranging from the cultural homoginisation associated with McDonald's fast food, to the industrialized agriculture associated with hormone-treated beef.
>
> (Heller 2002: 28)

We can see the overlap between the critique of 'malbouffe' and the critique of fast food articulated by the 'Slow Food' movement. Bové projected his own product, Roquefort cheese, as an aromatic metaphor for the notion of culture-as-nature (Heller 2002) which seems to conjoin social notions of traditional peasant life, culinary excellence of wine and cheese and also ecological good health. Bové also drinks copiously from the cultural anti-American feeling that is a key facet of French hostility to globalisation (Hay and Rosamund 2002). As if to hammer home the antithesis with Americanised agriculture Roquefort cheese is one of the items which the USA decided to penalise with high import tariffs in (WTO authorised) retaliation to the EU ban on USA (growth hormone treated) beef. Of course, much of Bové's (and, generally the green movement's) appeal to tradition can itself be labelled as a construction. Traditional peasant life was hardly idyllic. Moreover the way the organic industry is developing its well-researched, expert-informed holistic farm management systems and its effective marketing strategies, it is very modern in nature, as I have argued elsewhere (Toke 2002a).

We can see from this that opposition to GM food goes far beyond 'risk' politics (Hellier 2002), especially in France and Italy where hostility to GM food has become ever more intense over the past six years. The notion of 'quality before quantity' in food also influences many other parts of Europe. The USA can boast a thriving and growing organic agricultural sector that rivals the sizes (pro-rata) of organic sectors in several European countries. Yet it seems that the 'quality before quantity' approach to food is much more 'ghettoised' and cut off from the mainstream as opposed to Europe. Whereas the notion of becoming ecological is fashionable in some areas, such as parts of the Pacific North West of the USA, ecologically responsible living seems to be associated with 'downshifting', a critique of materialism. By contrast in Europe the Slow Food movement has appeal to 'haute cuisine' and is something to which many in the mainstream will aspire. In other words the notion of 'quality' food spreads much, much wider than supporters of organic food. For example, in the UK, city people moving into farm the countryside are seen, in a report in the *Sunday Times*, to be 'up' – rather than 'down'-shifting:

> Well-heeled townies who have swapped their brogues for wellies are now producing some of the best quality beef, pork, cheese and vegetables available in Britain ... Experts say their produce – which includes some rare and tasty breeds not seen for a generation – is helping drive a renaissance in British restaurant cuisine and home cooking.
>
> (Elliot and Iredale 2003)

These 'high quality' food products are expensive, and here lies a crucial nuance of difference between European and US attitudes to food. I was sometimes intrigued by the apparent consternation on the part of some US food industry spokespersons that the Europeans apparently cared so little about the cost savings in GM food. Now I am sure that many European shoppers do want cheap food, but the marketing problem is that among the European middle classes it is accepted that good food is often expensive. Ergo, cheap food may be junk food.

Of course, these cultural underpinnings which cast what is seen as 'malbouffe'/junk food in a bad light and which hamstring the notion of marketing GM food in Europe right from the start may not be the only explanation for European scepticism of GM food and crops. The climax of the BSE crisis occurred in March 1996 just at the time when GM food was about to be unloaded from US shipments. Even here it can be argued that the way that the BSE crisis was constructed, rather than the event itself, is significant. However there is a big case to be made that the BSE case was a major factor in triggering the European rejection of GM food and crops. I now want to turn to the arguments about the significance of the BSE crisis for European GM food and crop policy outcomes.

Was it BSE that did it?

Was the BSE crisis the principal cause of European rejection of GM food? I shall call this the 'Mad Cow thesis'. A positivist, event-based, analysis would seem to back this thesis, especially as it can be linked to behaviouralist studies (Petts *et al.* 2001: 35) such as that in the UK which links rejection of GM food with BSE crisis in the minds of the British public. Behaviouralist theory tends to be positivist in nature. The construction and interpretation of the data collected by behaviouralists (who make judgements about motivations according to what people say or do) is not an important part of these positivist approaches. The data is accepted as a 'given'. Certainly, analysts such as Vogel (2001: 19) declare that:

> It is impossible to exaggerate the significance of the regulatory failure associated with BSE on the attitude of the European public toward GM foods. This was especially true in Britain, where unfavourable press coverage of agrobiotechnology increased substantially following the BSE crisis: between 1996 and 1998 the percentage of those strongly opposing genetically modified foods rose from 29 per cent to 40 per cent. But its ramifications were felt throughout the EU.
>
> (Vogel 2001: 19)

I would definitely argue that it is possible to exaggerate the influence of the BSE, at least in Europe as a whole and even in the UK. It is plausible to argue that in the case of the UK that the BSE crisis reinforced fears of GM food. However, even though behaviouralist studies may pinpoint BSE-linked fears as the 'cause' of suspicion of GM food it is also possible that, in some people's minds, BSE became a new metaphor for an existing distrust of food safety and genetic modification. In Britain, as was discussed in Chapter 3, distrust of food safety and food quality had already manifested itself in earlier crises such as salmonella-in-eggs. Moreover, can we really be sure that the drift in opinion against GM food cited by Vogel was not at least partly caused by the fact that GM food only appeared in shops towards the end of 1996, and thus became a salient political issue? We do not know the answers to questions like these.

Now, I do think, as I discussed in Chapter 2 that we can discuss associations between different events and discourses, but we must be wary of attributing cause. We can have a discussion of what associations do or do not exist and thus make comments on the relative plausibility of particular explanations for events. It is in this context that I examine the plausibility of the 'Mad Cow' thesis explanation of European opposition to GM food.

Let us examine the circumstantial evidence that the BSE crisis was an important influence in harming the image of GM food in the EU. I shall now begin by considering the very considerable public controversy over imports of CIBA/GEIGY Bt maize (corn). I touch on this because it was the first major controversy (perhaps the crucial controversy) to afflict GM food and crop imports into the EU. By the way, CIBA/GEIGY later became Novartis which today trades under the name Syngenta.

The CIBA/GEIGY maize, you will remember from the previous chapter, contained not only recombinant genes for herbicide tolerance and insect resistance, but also an antibiotic resistant marker gene that was resistant to ampicillin. Again, you will recall from Chapter 4 that this caused very little comment in the USA when it was registered with the USDA and the FDA, and the issue of the ampicillin resistance appears not even to have been formally considered by the FDA when the product was de-regulated.

The actual arrival of Monsanto's GM soya in Europe came after the BSE crisis. However, the soya was authorised by the member states' GM food Regulatory Committee and the European Commission just before the BSE crisis erupted. The GM maize was therefore the first GM food product following the BSE crisis to be considered for EU approval by the Regulatory Committee (on GM crops) of member states. I shall describe the regulatory process at greater length later, but for the purposes of this story it is necessary to say that under the (Directive 90/220) legislation on deliberate release of GM products in force in 1996, a proposal for import (and cultivation) of GM seed had first to be considered by a member state. If the proposing member state (France in the case of the CIBA/GEIGY case) approved the proposal, in a 'dossier', then it was virtually a legal requirement that the Commission authorise the product. Now the dossiers are circulated for comments to member states who send representatives to a Regulatory Committee to discuss the dossier. If there are objections, the matter is also considered by the relevant Scientific Advisory Committee. However, the Regulatory Committee did not have (and its current equivalent still has not) the legal power to reject the proposal. That could only be done under the 90/220 Directive if Council of Ministers unanimously rejected the proposal.

In fact, when the Regulatory Committee met in June 1996, no less than 13 out of 15 of EU member states objected to the authorisation of the ampicillin-resistant GM maize (on the advice of their own, nationally based, food safety scientific committees). This vote only had advisory standing. As we saw in Chapter 3, the British (who were concerned about risks posed by the antibiotic marker resistant gene) were one of the 13 countries who objected. This left France, the proposer, as its only proponent apart from Spain. By contrast, the (Commission appointed) EU Scientific Committee for Food said that the maize did not present a threat to human health.

The support of the French government for the maize authorisation did not mean that French opinion was entirely quiescent concerning GM food. There is clear evidence of the invocation of mad cow disease as part of calls to beware of GM food. As two French analysts put it:

> On November 1st 1996, *Liberation,* one of the biggest French daily newspapers, headed its front page 'Alerte au Soja Fou' (Beware of the mad soyabean), in an article about the impending arrival of US imports of GM soyabean. By making a direct link between mad cow disease and GM crops, this provocative title anticipated a widespread public controversy about the place of the precautionary principle in the public decision making process.
>
> (Roy and Joly 2000: 248)

Greenpeace protests at the shipments occurred on both sides of the Atlantic as commercial pressure mounted on the European Commission to give a final go-ahead for the importation of the GM maize. Greenpeace argued that:

> Advice from the Commission's scientific committees did not predict the disastrous situation arising from BSE when animal carcasses were allowed to be used as animal feed. Let's hope the Commissioners will bear this in mind when they make their decision.
>
> (Greenpeace Press Release 1996)

Finally a decision was taken by the Commission on December 18th to allow the import and cultivation of the maize. The key issue which I wish to illustrate here is that it was the BSE metaphor that was used widely across the EU to oppose and criticise this decision. It is, at this point, worth quoting the first paragraph (translated into English) of the lead story in the Brussels based newspaper, *Le Soir.* This story occurred in the midst of threatened and real national bans on the importation of the maize and also in the context of a leaked document exposing the divisions inside the Commission on the subject. Emma Bonino, responsible for Consumer Affairs, criticised the speed in which the maize was authorised and she argued that it should be labelled. Andre Rich wrote:

> The Mad Cow lesson has not been learned! On 18th December, the European Commission authorized the entry of American GM maize onto the European market. According to confidential documents which we have been given, economic and political pressure has

prevailed over scientific and health precautions. Six months previously the Commissioners were divided over the issue of lifting the
embargo on British (beef) gelatine ... The document gives the
impression of an attempt to avoid scientific evidence ... What is even
worse, several Commissioners, especially Brittan and Bangemann
came out against (the idea of labelling) because they feared a conflict
with the WTO.

(Rich 1997a)

The decision to authorise the maize was later (in April 1997) deplored by
the European Parliament in almost visceral terms. The Parliament
demanded the suspension of the importation of the maize and deplored
the lack of attention to the precautionary principle, consumer health concerns and environmental issues. The resolution specifically warned of 'the
risks of transmission to humans of the antibiotic resistant marker gene ...
and the risks of use of pesticides' (Rich 1997b). The author of the *Le Soir*
article which reported the resolution commented that 'The mad cow
affair has worried everybody. This resolution is an important warning to
the Commission which is still tackling the Mad Cow Crisis' (Rich 1997b).

What is especially interesting about these responses to the CIBA/
GEIGY maize affair is that the argument that the Commission were ignoring scientific opinion in order to meet commercial pressures from the
USA is almost a mirror image of the US arguments. You will remember,
from the previous chapter, that the USA have alleged that the EU has
ignored the scientifically positive evaluation of GM food and crops in
favour of political pressures.

A wave of public reaction to the importation of the maize spread across
Europe, although (perhaps surprisingly in view of the Mad Cow thesis)
not so much in the UK where the reaction was fairly muted in the
1996–1997 period. In Austria pressure was mounting on the government
to ban the maize even before the Commission took a decision and two
leading supermarkets, Spar and Julius Meinl banned food containing GM
food from their shelves. Immediately following the Commission's decision
to authorise importation of the GM maize the Austrian government
decided to ban the import of the maize, citing the antibiotic resistant
marker gene as the chief reason (Netlink 1996). The Austrian decision
was followed by bans on GM maize by Luxembourg and Italy. There was
mounting pressure in other EU states, such as Sweden, where SABA, the
largest food chain also banished products containing GM food from their
shelves.

As time went on pressure grew on food stores to ensure that the products that were sold were GM-free. Labelling of food products containing
indentifiable (by presence of DNA) GMO ingredients was fully introduced
by September 1998. The process of supermarket shelves being emptied of
GM food products was quicker in countries like Germany, Sweden,

Denmark and Austria compared to places like the UK and Spain, but by the end of 1999 it was difficult to find labelled GM products on supermarket shelves anywhere in the EU.

It had been anticipated that a series of GM crop and food products would be approved. A small number trickled through the system in the 1996–1998 period, but opposition from some member states was intense. Finally, in October 1998 the whole process ground to a halt as the number of member states opposing new product authorisations grew and became more consistent. Only eighteen GMO products were approved during this period. Only five of these involved full approval for cultivation for food supply purposes (three herbicide tolerant oil seed rape, one Bt maize and one herbicide tolerant maize). What became known as a 'de facto' moratorium on new GM crop and food authorisations came into play. As I shall discuss in more detail later, the Commission felt it had no other choice but to permit this state of affairs. The recalcitrant member states demanded new legislation as their price for allowing the moratorium to be lifted. In general the new legislation that was required would step up the stringency of the GMO assessment procedure and also the criteria for labelling GM food. Again, I shall say more about this process in later sections since it is necessary for me now to focus on the arguments against the 'Mad Cow Thesis'.

Why it wasn't just BSE

There are several reasons why it would be very wrong to focus solely on the BSE crisis, or even recent food crises, as explanations for European hostility towards GM food. We have seen earlier, how European attitudes to food as a quality product differ significantly from dominant US attitudes. Hence it should be no surprise that there is a lot of evidence to show that the EU and its member states treated GM food and crops rather differently to the Americans even before the BSE crisis.

Essentially, the argument against seeing the BSE crisis as the major driver of differences between US and EU approaches to GM food and the more sceptical European attitude to GM food is that European scepticism to GM agricultural technology *preceded* the BSE crisis. The first piece of evidence is the clear difference in regulatory attitude between US and EU approaches that was manifested in the EU Directive 90/220 on Deliberate Release of GMOs into the environment. The prefix '90' indicates the year – 1990 – that the Directive was formally adopted, and this is several years before the BSE crisis. As I discussed in Chapter 3 (about the UK), GM plants are regarded as being fundamentally different to conventional plants. In the EU, specific laws have been created for assessing the release of GMOs whereas, in the USA, GM crops and food are regulated through use of pre-existing legislation.

The emergence of this pattern was not mere chance. As with all legislative proposals, they are first proposed by the European Commission, before being discussed by the relevant Council of Ministers and the European Parliament. There are two forms of major pieces of legislation; directives which leave various details of implementation up to the member state, and Regulations where precise details are mandated to apply equally to everybody across the EU. In the case of GMOs even the Directives on GM releases have been fairly prescriptive since the essence of the system is a safety assessment procedure set out in the Directives.

The European Commission's proposals for what later became the 90/220 Directive were first published in 1986. The Commission justified its proposals for separate legislation on the basis of three key influences. First, that new biotechnologies presented new risks, second on the basis of the EU's acceptance of what was emerging as the precautionary principle, and third:

> The Commission further justified its legislative intentions as a step towards creating a single market for GMOs, pre-empting disparities in national regulatory some of which (e.g. in Germany and Denmark) were looking increasingly prohibitive.
>
> (Shackley, Levidow and Tait 1992)

Now I do not want to conflate the way the Commission presented proposals with a definite cause for the Commission's position, but we can unscramble some possible pressures from the above statement. These were the pressures from greens, environmentalists and other dissident groups in Denmark and Germany, and other parts of Europe. Die Grunen had emerged as a new political force in Germany just at the time when proposals for work on GMOs were being put forward in the German state of Hesse in 1983. Greens mounted vigorous challenges to proposals by the pharmaceutical company Hoechst to build a pilot plant for the purposes of manufacturing insulin using genetically modified organisms. Indeed the campaign was successful in the sense that the plant's operation was delayed from 1983, when Hoechst's proposal initially emerged, to 1993 (Shackley 1993: 190–193). A leading Green-leaning academic (and critic of biotechnology), Erica Hickel, spoke of the need to 'establish, by democratic means the values that are to be held in science . . . the large technological corporations do whatever they want because they pretend they are doing value-free work' (Spretnak and Capra 1985: 109, cited by Shackley 1993: 188). Benny Haerlin, a German Green MEP and one of the foremost GMO-sceptics in the European Parliament criticised biotechnologists because they wanted 'to convert agriculture into a branch of industry' (Haerlin 1990, cited by Levidow *et al.* 1996: 137).

In Denmark the Environment and Gene Technology Act came into force in 1986. This involved lengthy procedures for consultation and detailed mechanisms for a range of institutions to make complaints about proposals to conduct genetic engineering. These procedures had the effect of delaying such industrial ventures. (Shackley 1993: 213–214). The German Genetic Engineering Act came into force in 1990. This was criticised by the biotechnology industry because of what were seen as heavy handed requirements for administrative authorisation to prevent environmental or health problems (Gill 1996).

A highly revealing piece of research by Shackley *et al.* (1992) discusses a battle within the European Commission between DGXII (Science) and DGXI (Environment) over the shape of biotechnology regulation. DGXII was more inclined to be sympathetic to the wishes of the biotechnology industry. On the other hand DGXI was more inclined to be sceptical of the influence of industrial corporations and sympathetic to pressures from environmentalist opinion. DGXII proposed, in effect, a minimalist piece of legislation applied to all novel organisms. In this scheme oversight of GMOs would be delegated to a regulatory body that was heavily influenced by the biotechnology industry. In response to this DGXI abandoned its desires for legislation covering all new organisms in favour of a more interventionist approach specifically aimed at regulating GMOs. Indeed the Environment Directorate allied with DGIII (responsible for single market harmonisation and industrial policy) to achieve this aim. In the process this overcame DGXII's desires for minimal GM regulation.

We can make an interesting comparison between the contending forces in the USA and the EU in the respective GMO regulation debates. In the 1980s we saw a battle, in the US Administration between the 'free marketeers' who opposed any different regulation of GMOs and the 'minimalist regulators' who argued (successfully) for a coordinated framework based on existing laws. Actors such as the Union of Concerned Scientists (UCS) and environmentalists who wanted a new law to cover GMO regulation failed to have their ideas considered by the Administration. By contrast, in the EU it was the 'free marketeers' who were the outsiders. The battle was between minimalist intervention and more rigorous (industry-sceptical) examination of proposals to release GMOs. The views of 'moderate' sceptics of GMOs (equivalent to US groups like the UCS) were, essentially, advanced by the Environment Directorate. However, unlike radical critics of GM technology, DG Environment still wished to see useful application of GM technology in agriculture. DG Environment's approach became the law. Indeed the legislative process, which involved what has always been the GM food-sceptical European Parliament, made the regulatory regime stricter. One analyst comments:

> Shortly after its enactment, Directive 90/220 came under attack by agrochemical multinational companies which were heavily investing plant biotechnology. According to them, it lacked a scientific basis, stigmatized GMOs and thus disadvantaged 'European' biotechnology.
>
> (Levidow 2001: 849, citing SAGB 1990)

The biggest point that arises from this discussion is that even in the 1980s we can see a distinct difference in the dominant approaches on GM technology in the EU compared to the USA. These differences increased quite considerably in the period leading up to the BSE crisis. We can also see from this discussion that Majone's notion (discussed in Chapter 2), that industrialists seek pan-EU regulation on environmental issues applies only to the concept of minimalist legislation, not the maximalist ideas often put forward by environmental and consumer groups.

A major fissure opened up between the approaches of the USA and the EU on genetic engineering technology through the rBST controversy. This controversy has resulted in a damaging trade conflict. rBST stands for 'recombinant bovine somatotrophin'. It is a substance, manufactured using genetic engineering, which is given to cattle to increase milk yields. The product was made by Monsanto, and after a battle with anti-GM activists, including Jeremy Rifkin's Foundation for Economic Trends, it was registered by the FDA for use in the USA. However, the application for use in the EU has never been authorised. The application was frozen in 1990, after which successive moratoria were imposed, followed by a permanent ban in 2000. The consequence has been that the EU has banned imports of beef from cows treated with rBST, which has effectively banned US beef exports to the EU. The USA took a complaint to the WTO which ruled in its favour. The WTO authorised the USA to impose trade sanctions on the EU, with French items such as Roquefort cheese taking the brunt of the sanctions. This encouraged José Bové, who I mentioned at the start of the chapter. He was already an energetic campaigner against US fast food and GM food and in the summer of 1999 he staged a demonstration involving the demolition of a McDonald's restaurant which was being built.

The significance here is that this is another example of differences between the USA and the EU on genetic engineering which is prior to the BSE crisis. Three factors have been adduced to explain differences between the USA and the EU on this matter (MacMillan 2002).

First, there were fears concerning animal health. Studies conducted by Monsanto revealed that cows treated with rBST were more likely to suffer from mastitis (inflammation of the udder). Monsanto said that this was because the cows were high yielding rather than because of the hormone itself, but others may regard this too fine a distinction to matter. Certainly the British Veterinary Advisory Committee recommended against authorisation, although the British government ignored this recommendation and in fact is the only EU member state to formally note its opposition to the EU's rBST embargo. This, incidentally, meant that the USA did not impose sanctions on British products.

A second reason was fears of impact on human health. This is a popular fear, and is one that overlaps with consumer concerns about the possible impact of GM food on human health.

Third was the fact that a crucial time frame for a decision on authorisation came directly following the decision by the FDA to register rBST in the USA. Unfortunately for Monsanto, the EU were, at that time (in 1993), grappling with the implementation of milk quotas for EU farmers in order to counter the problems of over-production of milk that had been affecting the EU's Common Agricultural Policy (CAP) for many years. The arrival of a product that would increase milk production still further was not exactly what EU decision-makers wanted or needed.

It is a matter of debate about the relative weight of these three factors, although the EU has tended to couch its opposition to rBST in terms of animal health concerns. It may be argued that this decision has a lot to do with farming policy rather than health or the environment, but even here the lack of European concern with the notion of efficiency and productivity in food production (compared to the USA) is something that does overlap with the debate about GM food and crops. The overlap here concerns the apparent relative European preference for food quality/style as opposed to low food price compared to dominant American sensitivities. There is also another connection between the rBST and the GM food/crop cases, besides the large fact that they both involve genetic engineering in agriculture. Some analysts in the USA fear that if the USA does win its case against the EU on the GM food and crops issue, then trade sanctions will, if the rBST case serves as an example, merely poison the political atmosphere further without helping US agricultural exports.

So far I have explained how the legal basis of GMO regulation, and the rBST dispute, illustrates how the USA and the EU diverged significantly on GM food and related issues before the BSE crisis. However, the case for this proposition does not end here. We can also see evidence which quite clearly shows how some EU legislation on labelling, a crucial area of distinction between US and EU regulatory policy, was already well on the way to being distinctive. Not only that, but also some member states had already signalled objections to EU authorisation of GM crops.

Labelling was to prove to be a watershed issue upon which the outcomes of the GM food crop outcomes turned. At this point it is helpful to say a little more about the EU legislative procedure. Once the Commission has formulated a legislative proposal for a Directive or regulation it goes to the Parliament for a first reading. It then passes on to the relevant Council of Ministers who produce what is called a 'common position'. This common position then goes to the Parliament for a second reading. Those amendments which receive an absolute majority of members of the European Parliament at this reading go forward to a 'conciliation' exercise whereby the Council needs to consider those amendments which are supported by the Commission.

An early sign of later battles on this crucial political battleground of labelling GM food came during the early debates on what later became Regulation No. 258/97 concerning 'Novel Foods and Novel Food

Ingredients' (Legislative Observatory 1992). Initially the (1992) Commission proposal contained no provision for labelling GM food. In addition, the Commission proposed a US-style simple notification procedure for authorising GM food products. GM foods would require no scientific assessment. All that would be required would be a statement that the food was substantially similar to its conventional equivalent. These proposals were strongly contested by the Environment Committee of the European Parliament who, at the First Reading in Autumn 1993, proposed a series of amendments. These aimed, among other things, to ensure that food containing GMO ingredients was labelled, that there should be a full safety assessment procedure for GM food products and that there should be public consultation about proposals to authorise GM foods. The full (plenary) European Parliament backed the idea of labelling GM food. In October 1995 this idea was controversially dropped by the Council of Ministers with the consequence that Germany, Austria, Denmark and Sweden voted against the Council's common position on the grounds that the measure did not contain adequate provision for labelling GM food.

Although the debate about labelling continued until November 1996 before being finally decided, we can see here that perhaps the single most important basis for the division between the USA and the EU, namely the labelling strategy, achieved the backing of the EU's legislative body nearly three years before the BSE crisis. It also received the backing of four member states in October 1995 who took the position of rejecting draft legislation that did not contain the provision for labelling.

The Environment Committee met in February 1996 to prepare for the Second Reading faced by a compromise 'Common Position' proposal on labelling that they found far too weak. (A Common Position proposal is the result of negotiation regarding the amended proposal between the Commission and the Council of Ministers). This Committee proposed amendments favouring labelling of all food containing genetically modified ingredients or enzymes. The Environment Committee also represented the demand for a full authorisation procedure involving scientific tests on the GM foodstuffs. Such tests would include tests on animals and also in vitro experiments on the effects on human blood and tissues. Although these demands did not have sufficient support to be carried into completed legislation at the time, they were to become the distinctive marks of legislation that is only now (at the time of writing) completing its legislative procedures.

The plenary session of the European Parliament met on 12th March 1996, eight days before the British government triggered the BSE crisis with its announcement that BSE had spread to humans. The Parliament adopted only six out of the 54 amendments to the Council's common position tabled by the Environment Committee with the required absolute majority of all members – a difficult requirement since usually only 70–80 per cent of MEPs attend. However, it did adopt some of the Environment

Committee's amendments, including the most crucial change. This involved mandating that GM foods containing recombinant DNA be labelled, whether or not they were 'significantly' different from conventional food. This amendment was contested by the Commission and the Council of Ministers, but the parliamentary representatives finally prevailed upon them to accept labelling on the basis of detectable recombinant DNA in November 1996. This was during the Parliament/Council Conciliation procedure.

The row about the importation of GM maize arose during and soon after this period, and of course the Novel Foods regulation did not come into force until May 1997. This left the (already approved) GM soya and maize unlabelled until a further Regulation (No. 1139/98) came into force in the Autumn of 1998 which required that food containing ingredients from these GM sources be labelled. Following controversies about 'accidental contamination' Regulation 49/2000 was introduced to exempt food from being labelled as GM food if the proportion of GM food is no more than 1 per cent.

It may be possible to argue that the occurrence of the BSE crisis may have finally tipped the argument in favour of labelling during the Conciliation procedure. I do not have, and most likely nobody has, the evidence to adjudicate one way or the other on this matter. It is reasonable to suggest that the BSE crisis and the CIBA/GEIGY maize affair may have hastened the labelling of GM soya and maize which had escaped the original Novel Foods Regulation. However, given the story that I have just described, the EU's move towards scepticism of GM food and crops looks much less like a conversion produced solely or even mainly by the BSE scare (or other ' food scares') and more like an event that was constructed in such a way as to reinforce an existing trend which already had considerable strength. As if to emphasise this point, the approval of the importation of Monsanto GM soya, which (as mentioned earlier) was taken just prior to the BSE crisis, was actually opposed by three countries; Austria, Denmark and Sweden – countries which also voted against the early draft of the Novel Food regulation for its failure to mandate labelling of GM food. Luxembourg abstained on the Monsanto soya decision.

What can be said with some certainty is that the Parliamentary (and member state) balance over whether derivatives of GM foodstuffs should be labelled has changed since the BSE crisis. Derivatives are processed products where there are no longer detectable traces of GM DNA. As a Commission official put it:

> The BSE crisis changed the majority in the EU. Before a minority wanted oil and starch labelled, but after the BSE crisis those wanting oil and starch labelled were in a majority.
>
> (Interview with DG Environment Officials – 2003)

We can see that the role of the BSE crisis, though important, is rather more limited than some accounts would imply. As I argued in Chapter 3, when people cite the BSE crisis as a cause for being worried about GM food they may be using a resonant metaphor to highlight a concern that would have existed anyway. Positivist accounts (or accounts that are implicitly positivist by their unquestioning assumption that events cause certain outcomes) will not focus on such interpretive explanations. It seems plausible that the BSE crisis reinforced existing concerns, and increased criticisms of GM food in states such as the UK, but we cannot even be certain of the strength of this linkage. This is because we cannot be certain that concern about GM food increased in places like the UK simply because it happened to be that GM food imports arrived from the USA in late 1996. To some extent, the apparent link with the BSE crisis, which also occurred in 1996, may just have been a coincidence linked by little more than that they were both contemporary examples of European concern about food quality issues.

Another event-related (or, in this case non-event) explanation for EU concern sometimes cited include the fact that labelling was not intro-duced until after controversy had erupted. As mentioned in Chapter 2, some argue that the failure to label GM food from the start of their impor-tation into the EU is an explanation for the outbreak of concern. Yet this fails to explain how the EU case is so different from that of the USA where there is no mandatory labelling of GM food. Moreover, even if the Oregon poll is even vaguely reflective of US opinion, there is much less demand for labelling of GM food compared to the demand for labelling in the EU.

This critique of the notion that the main differences between the USA and the EU over GM food and crops can be ascribed to the BSE crisis is a cautionary note to those who seek 'cause and effect' explanations for political outcomes. It also acts as an empirical example of the theoretical criticisms (made in Chapter 2) of advocacy coalition theory which relies too heavily on the notion that events change ideas and change the balance of power between different coalitions of interest. It is plausible to assume that to some extent the BSE crisis can be identified as being associ-ated with an increase in European hostility to GM food and crops. However, the interpretation of the BSE crisis as a problem for GM food may itself be conditional on a pre-existing attitude to food and environ-mental regulation that is different to that which holds sway in the USA. Now this does not mean that the social sciences can say nothing about understanding how outcomes occur. However we must produce an under-standing of both events and context and we must be wary of assuming that the interpretation of events is unproblematic, especially in a comparative study such as this. I will now structure some discussion around key regula-tory discourses concerning the frameworks for scientific regulation of GM crops.

Regulating GMOs – earlier rules

It makes sense to follow the practice employed in Chapters 3 and 4 in terms of dividing up the earlier regulatory discourses (up until 1998) from later regulatory changes. I have already discussed some of the background to the development of the 90/220 Directive on Deliberate Release into the Environment of Genetically Modified Organisms. I should point out that accompanying the 90/220 Directive was a Directive on contained use of GMOs which is numbered Directive 90/219. However, contained use has rarely been an item of controversy in the 1990s, so I will not discuss it further. That is, except to comment that anti-GM activists these days propose that GM experiments be conducted in contained circumstances. In the 1980s they tended to complain about contained use. This, along with the acceptance of medical uses of biotechnology is another example of how there have been significant shifts in radical green attitudes towards genetic engineering. I now want to look at some of the distinctive detail of the nature, and debates surrounding, the EU Deliberate Release legislation.

As I discussed earlier, the 90/220 Directive was formulated in a European atmosphere whereby environmentalist discourses of distrust of biotechnology industries had achieved widespread resonance. Such discourses manifested themselves even in statements made by DG Environment during the preparation of Commission proposals for the 90/220 Directive. For example, a joint DGXI/DGIII document stated:

> A lack of candour on the part of some companies about the potential environmental risks from their products coupled with a bland attitude of 'we know best' on the part of scientists and industrialists could pull the rug out from under these industries.
>
> (CEC 1986, cited by Shackley *et al.* 1992)

This approach is to be sharply contrasted with the 'free market' discourse which underpins the US regulatory discourse on GM food and crops. The very existence of the Directive is a direct contrast with the US approach which specifically rules out separate legislation for regulation of GMOs. Indeed, the existence of the US 'Coordinated Framework' itself (as opposed to the FDA's original preference for exclusive reliance on existing laws) seems to flow from controversies over the release of microorganisms, and the court-imposed moratorium on such an experiment (see Chapter 4 for details on this). However, in the EU the anti-GM movement succeeded in influencing legislation on genetic engineering as a whole, and initially even medical uses were controversial. Danish and German legislators felt obliged to adopt stringent gene technology laws involving extensive controls on industry. We can see here that the relative degree of intervention and breadth of that intervention in the USA and the EU seems closely related to the degree and type of public resonance with environmentalist GM-sceptical discourses that was in evidence in the EU and the USA in the 1980s.

We can see evidence of the more limited US concern with GMO regulation in the very definition of GMO. The US Coordinated Framework defined the regulated objects in terms of them involving 'pathogens', which was a roundabout way of pointing to GMOs since this referred to the process by which the DNA was transferred rather than the recombinant DNA itself. By contrast, Article 2 (2) of the 90/220 EU Directive distinguishes GMOs from what occurs 'naturally': '"Genetically modified organism (GMO)" means an organism in which the genetic material has been altered in a way that does not occur naturally by mating and/or natural recombination' (Official Journal of the European Communities 1990a). This is perhaps the most fundamental discursive difference of them all. Are GM plants 'natural'? The USA says yes, the EU says no.

Directive 90/220 contained three important procedural aspects that were different to the procedures inherent in the US system of GM crop regulation. The EU system demands assessments of proposed releases at two stages; first at the stage of applying for permission to conduct trials (Part B application), and also at the stage of applying for permission to commercialise the product (Part C application). Under the US system (as described in the last chapter) the safety assessment was considered when permits for trials are requested, although in practice the large majority have been dealt with through simple notification of intent rather than assessment. Commercialisation takes place after a permit is requested, and then granted, for deregulated status. Perhaps the best way of comparing the EU's system is to say that in the US system the product put up for commercialisation is 'innocent until proven guilty', whereas under the EU's system the balance was placed the other way around; it has to prove its innocence. In effect, the product had to be proved safe at two stages.

A second important difference is the residual powers of the member states under 'Article 16' of the 90/220 Directive. Although the preamble to the Directive 90/220 rules out the possibility of member states to 'prohibit, restrict or impede the deliberate release of a product ... properly authorized under this Directive', Article 16 para 1 states:

> Where a member state has justifiable reasons to consider that a product which has been properly notified and has received written consent under this Directive constitutes a risk to human health or the environment, it may provisionally restrict or prohibit the use and/or sale of that product on its territory. It shall immediately inform the Commission and the other Member States of such action and give reasons for its decision.
>
> (Official Journal of the European Communities 1990a)

In the USA, individual states have always deferred to the Federal agencies on issues of GM crop safety. A third crucial difference was the so-called 'Article 21' Regulatory Committee. This consisted of representatives of member states who discuss dossiers (scientific assessment details) of proposals for product release authorisations. In fact the device as set out in the Directive was, in formal terms, only a consultative nicety for the benefit of the member states. This is because even if the Regulatory Committee objected to the product proposal (as happened in the GM maize and various later cases), Article 21 of the 90/2220 Directive stated that if the Council of Ministers failed to act then the proposed measures would be adopted by the Commission within three months as a matter of course. The Council could only (legally) stop a proposal by a unanimous vote. However, since a member state was acting as 'raporteur' in the first place, the consultative actions of the Article 21 Committee could, in theory, do nothing to stop a proposal. In practice it became a different matter. Despite the nominal legal authority of the Commission, the fact that a coalition of objecting countries could delay proposals at the 'Article 21' Regulatory Committee was a clearly visible factor in frustrating EU GMO authorisations. Equally, the fact that individual states had powers to 'ban' GMO products under 'Article 16' also had effects. I shall expand on these issues when I come to examine the role of the Commission in more detail.

There was one way in which, in democratic terms, Directive 90/220 lagged behind the US system of GMO regulation. This was in the area of public consultation. As I discussed in Chapter 4, the US system involves a system of public consultation as a matter of form, but this was not the case with the European system of scientific regulation under Directive 90/220. Despite the efforts of the European Parliament to insert clauses making public consultation on proposals for GMO authorisations compulsory, the best that was achieved was Article 7 which stated that 'Where a Member State considers it appropriate, it may provide that groups or the public shall be consulted on any aspect of the proposed deliberate release' (Official Journal of the European Communities 1997).

The Novel Foods and Novel Food Ingredients Regulation, approved in 1997 (with amendments in 1998 and 2000) dealt with novel foods in general, including food derived from or consisting of GMOs. If the GMOs were 'live', that is they could reproduce, then they had to be authorised under the Deliberate Release Directive, but if not then they would escape the need for full risk assessment including tests and experiments. Notification of information attesting to the safety of the food was still mandatory, as opposed to the voluntary nature of the notification required by the FDA. However, at this point the food assessment procedures had a lot in common in that safety assessments were both based on the notion of 'substantial equivalence'. This means that if the food could be demonstrated to be the same, in nutritional terms, as its conventional counterparts, then it was assumed that it was safe. This principle was attacked by GM-sceptics in both the EU and the USA.

The big difference with the US system, in this original Novel Foods Regulation, was that if recombinant DNA could be detected in the food then the food had to be labelled as having been produced from GMOs. As was discussed earlier, the European Parliament was instrumental in changing the wording of what became Article 8 (4) of the Novel Foods Regulation. The criteria for labelling hinged on whether an analysis of 'existing data can demonstrate that the characteristics assessed are different in comparison with a conventional food or food ingredient' (Official Journal of the European Communities 1997). The word 'significantly' which came before 'different', in the draft agreed by the Council and the Commission, was deleted after Parliamentary amendment. On the other hand, attempts to have all food derived from GMOs labelled as produced from GMOs regardless of whether DNA could be detected were, at that stage, defeated.

In 1995 a Danish annex to an earlier (and for the Parliament, less satisfactory) version of the Regulation stated that consumers had to be given 'freedom of choice' through GM labelling. 'Only information which is sufficient to inspire consumer confidence will incite consumer acceptance of the new products and new technologies' (Legal Observatory 1992). In fact, for the opponents of GM food and crops, labelling was a form of power to limit GM food sales. Just as Foucault talked about dominant discourses being imposed on the subject prisoner through surveillance in prison, so the discourse on 'freedom of choice' for the consumer used labelling as a form of surveillance. This surveillance produced knowledge about the prevalence of GM food, and it is the act of producing knowledge that is synonymous with a shift in power relations towards those who are sceptical of GM food. This is an example of what Foucault meant by power going hand-in-hand with knowledge (Foucault 1977, 1980).

On the other hand rational choice institutionalists, who see incentive structures such as labelling regimes as a means of altering an actor's behaviour, might point out that in the context of pressure from anti-GM consumers supermarkets had an incentive to clear GM food from their shelves. As the 1998 deadline approached for labelling of GM soya and maize, GM foods disappeared off shelves in the UK. Today, most EU supermarkets obey a simple policy of not selling anything with a 'made from GMO' label. On the other hand GM food had already disappeared from shelves in countries like Denmark, Austria and Germany even before labelling took place. Perhaps the lesson here is that institutions operate partly through incentive structures but also partly through the belief systems that constitute the institutions in the first place. The problem with rational choice institutionalist accounts is that they do not seek to explain the emergence of the belief system upon which the institutional incentive structure is based. This is something that can be done through discourse analysis. We can, for example (as was done in Chapter 3 in the case of the UK) analyse how changing belief systems have coincided with changes in action by government at the national level. Foucault's 'power/knowledge' approach links study of the discourses with associated practices and use of particular knowledge/surveillance systems whose legitimacy is sanctioned by the dominant discourses.

I commented in Chapter 4 that rational choice theorists might try and predict outcomes through modelling 'games' where farmers choose whether to support labelling of GM food. I suggested that this type of analysis, while having uses, was limited in its central project of predicting possible outcomes since there was uncertainty how different actors in the 'game' would interpret their self interests. Pro-biotechnology actors have accepted the principle of labelling in the EU, although it is damned in the USA. As a representative of Bayer (which bought up Aventis Crop Science) put it:

> The situation is totally different in North America, US and Canada and Europe. It's true that it is challenging for Biotech companies to have a global position on GM labelling. In Bayer we simply recognise that in different parts of the world, there are different political positions, different approaches to GM labelling and it's true that our position is not the same in North America ... we adapt our position on GM labelling to local conditions, local context.
>
> (Plan interview – 2003)

Judgements about scientific uncertainty are belief systems rather than incentive structures and are difficult to 'predict' using game theory. The belief systems are an important point of comparison between the USA and the EU over GM food and crop policy. The EU Scientific Committee on Food's (SCF) opinion of the controversy about antibiotic resistant marker genes in the 1996–1997 period coincided with the US view on this topic. The role of the scientific committees (who, as I shall discuss later, are appointed by the Commission) is to give opinions when member states have objected to proposed EU product authorisations or about whether member states are right in refusing to authorise products on their territory. The Scientific Committee on Food (SCF) responded to the Austrian refusal to allow imports of the CIBA/GEIGY maize (which contained an ampicillin-resistant market gene) by upholding its existing conclusion that:

> The possibility that the product would significantly add to the already widespread occurrence of ampicillin resistant bacteria in animals and man is remote ... Thus the Austrian information does not cause the SCF to consider that the CIBA/GEIGY Maize constitutes a risk to human health.
>
> (SCF 1997)

Nevertheless, the SCF's views were at variance with the views of most of the food safety advisory committees of the member states, as I discussed earlier. Whether or not we take the views of the SCF on the CIBA/GEIGY case into account, we can see that by 1997 regulatory discourses in the EU were already distinctive to those that are still dominant in the USA today. We can also see that the Commission and its scientific advisers were fighting against efforts to increase these differences. This proved to be a losing battle. Even by this date, the differences with the USA (particularly on labelling) were already enough to trigger serious trade arguments. The situation was to deteriorate as dominant discourses in the EU, as expressed in legislation, treated agricultural biotechnology with greater scepticism.

Ironically, one vector for exacerbating differences between the USA and the EU, Directive 18/01, was originally conceived as a measure that would increase the similarity of EU and US GM crop regulation. In 1996, when the dominant discourse at the European Commission was that many GM crop products represented 'low risk' cases, the Commission produced proposals to streamline the system of scientific assessment:

> Experience has shown that the (GMO assessment) procedure followed is difficult to implement, time consuming, and cumbersome to follow both for users and authorities ... It has to be remembered that in this fast-moving, high-tech field, not only future notifications but also current ones already concern products similar to authorized ones in the EU, as well as products which themselves have been used elsewhere in the world and which have proven to be safe ... There is therefore a need to provide a streamlined procedure for those products posing no, negligible or low risk.
>
> (CEC 1996: 8)

The Commission's discourse on GM crop safety wanted greater regulatory sympathy to be shown to the agricultural biotechnologists. However this discourse did not resonate with large sections of EU public opinion. In fact, by the time that the revised Directive 18/01 appeared on the statute books it had been amended so as to increase, rather than reduce, the regulatory scepticism towards GMO releases.

Changing regulatory discourses

Directive 18/01 replaced Directive 90/220 and is significant for at least six key changes. First was the insertion of the 'precautionary' principle (Preamble (7) and Article 4 (1)) on top of the (re-stated) commitment on 'preventive' action to protect the environment.

Second was a new Preamble point (9):

Respect for ethical principles recognized in a Member State is particularly important. Member States may take into consideration ethical aspects when GMOs are deliberately released or placed on the market as or in products.

(Official Journal of the European Communities 2001)

This has considerable implications for developments such as genetically modified animals. Indeed it is difficult to see how this sort of development is going to be commercialised in the EU, certainly for food production. Medical uses may be different, although xenotransplantation is controversial in itself.

Third, were some restrictions on use of antibiotic resistant marker genes. Article 4 (2) stated:

Member States and the Commission shall ensure that GMOs which contain genes expressing resistance to antibiotics in use for medical or veterinary treatment are taken into particular consideration when carrying out an environmental risk assessment, with a view to identifying and phasing out antibiotic resistance markers in GMOs which may have adverse effects on human health and the environment.

(Official Journal of the European Communities 2001)

The use of these types of marker genes was to be phased out in products given full authorisation by 2004. This is very different to the USA where there is nothing more than rather vaguely worded draft voluntary guidance on the issue, as I discussed in Chapter 4.

The fourth difference was the inclusion of mandatory public consultation concerning product proposals, at the member state level (Article 9) and the European Commission (Article 24). This meant that the EU now gives more, rather than less, information on GM crop trials compared to the USA, since it is now normal practice in EU member states to give locations of GM trial sites. This type of information is kept secret by the United States Department of Agriculture.

The fifth difference, if it can be put in one bag, is regarded by most observers as the most substantial. It concerns the criteria for environmental assessments of GMOs. The 'Principles for Environmental Risk Assessment', contained in the Annex of the Directive 18/01 take up around eight pages of the Official Journal of the European Communities compared to barely one and a half pages in the 90/220 Directive. Among the new criteria laid down were extensive requirements for information of effects on non-target species, information on competitive advantages that may be transferred to other plants and information on the wider impact on ecosystems, including the food supply for birds and other animals. Trials need to be conducted in order to discover such information. Moreover consents for GMO products would last only for seven years (without being renewed), during which time post-release monitoring of environmental impacts would take place.

A sixth difference was a change in the duties of the Commission and the Council of Ministers regarding GMO product authorisations. The Council of Ministers had, under the 90/220 Directive, to be unanimous before it could reject a dossier; an impossibility given that there would not be a proposal without at least one 'raporteur' country backing it. Now, under the 18/01 Directive it became possible for the Council to reject a dossier by a qualified majority. There was also a tightening of the criteria under which individual countries could ban products, in an effort to dissuade member states from acting unilaterally.

It remains to be seen how effective either of these provisions are. Many would argue that the last-minute amendment to the 2003 Food and Feed Regulation (which will be discussed later) giving discretion to individual member states to take action to stop 'contamination' of conventional and organic crops more or less completely undercuts the ability of the Commission to stop member states taking unilateral action against GMO authorisations.

We have (in both the 90/220 and 18/01 Directives) a 'two' level system of regulation whereby GMO product proposals are assessed. First, they are assessed at a nation state level, and second at a supranational level by other member states through the Regulatory Committee and, in the event of a dispute, through the Scientific Advisory Committees. Using Scott's (1995) notion of 'regulative' rules, or the notion of incentive structures as put by rational choice institutionalists, we can see that these structures may have an independent effect on outcomes. Yet it is difficult to disentangle these effects from the effects of belief systems at a national state level, belief systems or dominant discourses which have become hostile to GM food and crops in many EU member states.

The passage of the 18/01 Directive was not enough for the six states who supported the moratorium on GM food and crop authorisations (seven states including Belgium from 2001). They wanted, among other things, new rules on food and feed authorisation and also on traceability and labelling to fill in what the GM-sceptics saw as a lack of scientific assessment and consumer information. These rules were, after a lengthy and highly controversial series of debates, adopted as Regulations 1829/2003 (on GM Food and Feed) and 1830/2003 (on Labelling and Traceability of GMOs and GM Food and Feed products). The Regulations were processed and adopted simultaneously.

This legislation involves the abandonment of the notion of 'substantial equivalence' as a means of safety assessment. First, the fullest information on the characterisation of all DNA changes has to be presented for assessment. Second, not only has information to be given but (as mentioned earlier) a specified range of feeding tests on animals and also 'in vitro' experiments must be conducted to study the practical biological effects of the GM food products. It is not this aspect that tends to worry the biotechnological or food manufacturers unduly. What does worry them is the fact that the GM food must be labelled on the basis of whether something is derived from GMOs irrespective of whether recombinant DNA is still present. Hence, in the original proposal for the Food and Feed regulation, the 'Whereas' preamble (20) states:

> The labeling should include objective information that a food or feed consists of, or contains or is produced from GMOs; clear labeling, irrespective of the detectability of DNA or protein resulting from the genetic modification in the final product, meets the demands expressed in numerous surveys by a large majority of consumers, facilitiates informed choice and precludes potential misleading of consumers as regards methods of manufacture or production.
>
> (CEC 2001)

Again, we see a discourse concerning the requirements of consumers' surveillance (or in rational choice language, incentive structures) in the form of labelling. This time (in Regulations 1829/2003 and 1830/2003) the surveillance depends not merely on detection of DNA but also, in the case of derived oils, starch and other products, on a paper traceability system connecting a given derived product, ultimately, with a given crop source. In addition, for the first time, all GM animal feed will be labelled. This will make it easier for supermarkets to demand that meat sold in their store be reared on non-GM animal feed. Given that animal feed and derived food products now constitutes the market for most of the remaining US GM soya used in the EU, it is reasonable (as discussed in Chapter 4) to assume that the sale of US soya to the EU to supply both animal feed and production of derived foodstuffs will continue to decline. The same criteria also apply to farm produce from any other parts of the world. This puts extra pressure on farmers around the world not to allow GM crops (mainly maize, soya and oil seed rape) to be grown in their regions, and also for GM food to be labelled in their countries so that it can be segregated, traced and labelled for export to the EU.

The GM Food and Feed and Traceability and Labelling rules were given a third reading by the European Parliament in July 2003. As this legislation was being processed through the EU machinery a parallel discussion was going on about 'coexistence' of GM and conventional farming. Anti-GM activists have campaigned for strict coexistence rules involving large crop separation distances to prevent 'contamination' of conventional and organic crops. The Commission initially tried to resist such blandishments, but eventually agreed to compromise, in the Food and Feed Regulation

1829/2003, for a formula that gave discretion to national governments to stop such contamination. The compromise wording says, at one point that: 'Member states may take appropriate measures to avoid the unintended presence of GMOs in other products', but in the next clause there is a statement that 'The Commission shall . . . develop guidelines on the coexistence of genetically modified, conventional and organic crops' (Official Journal of the European Union 2003: 21). This was, as the cliché goes, classic Euro fudge which has led to the emergence of several disputes between on the one hand, the Commission, and on the other hand, anti-GM food regions and member states, about whether GM crops can be banned. Then again, there is not much chance of them being grown in such places (or most of the EU) anyway, because of lack of consumer demand for GM food.

I have now sketched out some of the key discourses which make up what I would call (according to the theory in Chapter 2) the 'discourse' of the EU's discourse paradigm on GM food and crops. I have still to examine how the EU's scientific advisory system fits in (or does not fit in) with this approach.

Scientific advice on GM food and crops

In the previous two chapters I deployed the notion of the 'discourse-paradigm'. I did this as a way of conceptualising:

1   How judgements about what are seen to be scientific issues relating to GM food and crops are done in relation to a distinctive set of regulatory discourses which are linked to culturally based attitudes.
2   How scientists, who are informed by these commonly held cultural attitudes, are imbued with a distinctive set of attitudes about issues of controversy on agricultural biotechnology.
3   How these sets of attitudes produce a distinctive set of 'facts' which needs to be collected.
4   How the scientists involved in scientific advisory work are broadly representative of attitudes to GM food and crops held by their respective national scientific 'peak' organisations, in these cases the National Academy of Sciences (NAS) in the USA and the Royal Society in the case of the UK.

I set out, in Chapter 2, how we may adapt Foucauldian and Kuhnian notions of discourse and paradigms to describe circumstances which are associated with the conceptualisations that I have just outlined. A discourse-paradigm is associated with a set of scientists whose ideas are drawn from a distinctive national cultural background.

Extending this notion to the EU is more problematic, not so much because of any inadequacies in the discourse-paradigm notion as applied to nationally based cultural-scientific systems, but because the EU is not a country! It does not have a commonly experienced set of cultural attitudes in the way that is normally enjoyed by a nation state. Moreover there is no pan-European equivalent to the National Academy of Sciences or the Royal Society.

Nevertheless I do believe we can make effective use of the notion here, partly as a way of contrasting the differences between the role of scientific advice on agricultural biotechnology with that which obtains in EU member states and those which emanate from the EU's scientific advisory structures. Partly, also, we can see how the dominant EU regulatory discourses on GM food and crops which I have discussed earlier are a sort of composite of EU dominant discourses on GM food and crops, a composite that has been shaped by the European Parliament. However, as we shall see, the scientific advisory network is a creature of the European Commission. Hence there is no clear way that scientists in these discourses are imbued by this composite discourse in the way that, say, British scientists, are influenced by prevailing British notions of what it means to conserve wildlife (though bird protection).

The EU Scientific Advisory system has, in recent years, been a source of controversy owing to its perceived failure to urge effective action over the BSE dispute (Randall 2001). Randall explained the conclusions of a European Parliamentary Report (Chaired by Medina Ortega) on the BSE affair as follows:

> Mr. Medina Ortega and his colleagues were in no doubt that if the Commission had had a properly organised and independent machinery for multi-disciplinary advisory committees on public health related issues it would have stood a much better chance of making a more timely assessment of the evolution of the BSE epidemic and identifying the public health risks associated with it.
>
> (Randall 2001)

The EU's scientific advisory system was reformed in 1997. There were a number of changes, but three were especially significant to this discussion. First, the nature of the Committees was changed. Prior to November 1997 there were six committees covering animal nutrition, veterinary issues, pesticides, and also toxicology and ecotoxicology. Following the post-BSE reforms some of the committees remained the same, but the pesticides and toxicology and ecotoxicology committees were abolished and replaced by the Scientific Committee on Plants. The second key change was that the procedure for appointing members of the scientific advisory committees was made 'transparent' through competitively advertised appointments of the most qualified people for the job as opposed to the previous system which relied on members being nominated by member states. The third change with significance for this discussion was that the advisory committees were made answerable to the DG for Health and Consumer Affairs rather than the subject-relevant

Directorates where industrial interests could wield influence that may be inappropriate.

Most recently, the advisory committees have been made responsible to the European Food Safety Authority (EFSA) which has been established under Regulation 178/2002. The EFSA sounds like a copy of the US Food and Drug Administration (FDA), but in fact, as critics contend, it is a pale reflection since it merely gives scientific advice rather than being able to issue rules and enforcements as is the case with the FDA. A similar criticism is made in relation to domestic, allegedly 'toothless', bodies like the UK's Food Standards Agency which was established in 2000.

Although, under the post-1997 rules, scientific advisory committee members have to declare any interests before discussions of specific proposals, it is still quite acceptable to be a member of an advisory committee on GM products whilst having contracts with private sector biotechnology companies. In this sense, the EU practice is different from the rules which now obtain in the Scientific Advisory Panel (SAP) system in the USA and the Advisory Committee on Releases into the Environment (ACRE) in the UK, as I discussed in the previous two chapters.

In the EU, three Advisory Committees are heavily involved with giving opinions on GM crop and food products, the Scientific Committee on Plants (SCP), the Scientific Committee on Food (SCF) and the Scientific Committee on Animal Nutrition (SCAN). The Committees often work collaboratively on specific topics and their concerns overlap to a large degree. A plant biochemist who has worked on GM crop research for a number of years, and who sits on the SCP, criticised the notion of barring scientists who have relevant commercial interests from membership of scientific advisory committees:

> What such a rule is saying is that we don't trust scientists. The truth is that it's usually the best scientists that companies wish to recruit as consultants or otherwise and we need the expertise of such scientists for a range of purposes. At any SCP meeting we have to state whether we have any conflicting interests. At the end of the day it is whether you trust people to do the job. You can have bias without having any commercial interests. Political pressure for changes to the ACRE panel appear to be stronger than we have seen in Brussels.
>
> (SCP scientist interview – 2002)

The formal role of the EU scientific committees is to give opinions when member states object to proposals for product authorisations. In the case of applications for GM crop authorisations this has happened every time, except in the case of carnations modified for colour. These objections usually take place after the raporteur (the country dealing with the dossier) has circulated other member states under the 'Part C' process for commercial authorisation. However, another important function of the advisory committees has been to produce opinions on actions by individual states to ban or restrict particular product authorisations in their territory. In almost all cases the SCP and the SCF have ruled against the objections. In particular the SCF ruled against objections made about CIBA/GEIGY maize before its original authorisation in 1996 and, as I have noted earlier, in 1997 on the occasion of Austria's refusal to allow the product to be placed on the market.

There is just one case where the SCP has given a negative verdict on a proposed authorisation. This was in regard to a Dutch proposal to authorise a 'high amylopectin potato cultivars' product from Avebe. The SCP concluded, in October 1998, that there had been 'insufficient risk assessment', particularly on an antibiotic resistant marker gene 'which confers resistance to amikacin, a clinically important antibiotic' (SCP 1998).

In their scepticism of the Adebe potato application the SCP showed a sensitivity to the strong swell of opinion against giving authorisations to products containing antibiotic resistant marker genes. However, the SCP has not shown any degree of sympathy with objections, cited at the member state level, to GM crops on grounds of the spread of herbicide resistance, the spread of insect resistance to Bt toxin or harm to non-target insects by products containing the Bt toxin. The SCP has not shown any wish to delay products on the grounds that assessments on their impact on the food supply for birds had not been carried out. This is something upon which the UK government's reformed (ACRE) has insisted.

The membership of the SCP is, in terms of academic scientific discipline, quite different to the reformed ACRE that I studied in Chapter 3. I discussed how the majority of the membership of ACRE is now concerned with ecology, plants and insects. This is different in comparison to the SCP membership. As far as I can judge by my own analysis using the biographies of the SCP posted on the EU SCP website, the large majority of the SCP are not actually experts on plants themselves, but on various types of human and animal toxicology. This membership betrays the SCP's history as a body that has been formed following the merging and renaming of the pesticides and toxicology/ecotoxicology committees. It does strike me as a body that is ideally set up for consideration of pesticides rather than GMOs. Four out of the 18 members of the SCP are specialists in biochemistry and biotechnology, four specialise in plant ecotoxicology and ecology, seven in human toxicology and three in veterinary-related animal toxicology.

One has to ask whether this is an optimum arrangement. I must venture the opinion that either the committee should be reorganised as is the case with the UK's Advisory Committee on Releases into the Environment (ACRE) to take account of what are, ultimately, politically and culturally determined regulatory scientific priorities or there ought to be a switch to the US system of scientific advice whereby expert panels are selected as appropriate to particular issues.

Members of the SCP have seen 'themselves as protecting scientific risk assessment from political bias – by contrast to national regulatory procedures, which have been influenced by anti-biotechnology pressure groups' (Levidow 2001: 851). An SCP member is quoted as saying: 'We are asked only scientific questions. The definition of adverse effects is not a political question – only a scientific question' (Levidow 2001: 851).

Judgements about what is or is not a relevant scientific issue which needs to be researched has changed. For example, since 1998 there has been more emphasis on the impact of herbicide tolerant GM regimes on food for wildlife. There has also been concern about the spread of herbicide tolerance to weeds. A member of the SCP commented that:

> Risk assessment is a continually evolving process as more scientific information comes online. The problem is the baseline against which to compare any impact of the GM crop (traditional agriculture, organic agriculture and over what time period etc) ... I think that there will be several challenges facing risk assessment in the future. One of them is gene stacking. The Scientific Committee has not had to deal with this question specifically as yet. The next generation of GMOs with stacked traits are going to offer more of a challenge to the risk assessors.

> (SCP scientist interview – 2002)

A rational choice analyst may put this in terms of 'new' information becoming available. However, an interpretive analysis would argue that the scientists were displaying the effects of reacting to changing dominant discourses on what constitutes culturally acceptable environmental risks and responses to scientific uncertainty. As I have previously discussed, the distinction between what is 'political' and what is 'scientific' is itself doubtful. This confusion coincides with the sort of positivist error which I discussed in Chapter 2. Opinions about uncertainty maybe called scientific, but then what is called science is culturally constituted and hence cannot in such instances be reduced to positivistic assessments. The notion that the experts somehow pursue a sort of non-political function is therefore fallacious.

It does seem that in terms of their general attitude to scientific assessments of GM food and crops the EU scientific committees are somewhere in between what may be characterised as a US discourse-paradigm and a typical European national discourse-paradigm. It is difficult to say that the EU scientific advisory committees form part of a discourse-paradigm themselves. They seem to be far too much in a vacuum with their only atmosphere provided by the appointments procedure of the European Commission and a second hand influence of European public opinion. The advisory committees must be regarded as servants of the Commission itself, and somewhat removed from the evolving priorities in the European Union as represented by dominant discourses thrown up by decisions of the European Parliament in particular.

This is a useful launch point for an examination of the relationships between the Council of Ministers, the European Commission and the European Parliament that have been exhibited by the GM food and crops issue.

Institutional theory and the EU

In Chapter 2 I set down a section of institutional theory that I wanted to specifically apply to the EU in order to both examine the relevance of the theory itself and also to use the theory to explain outcomes on the GM food and crops issue. I have already discussed the relevance of Majone's (1992) injunctions on environmental regulations. Then, there is the issue of the theory of 'functionalism'. As I discussed in Chapter 2, functionalist theory suggests that technical measures of EU harmonisation would lead to 'spillover' at the political level. This has happened in the case of GM crop and food regulation to the extent that the rule-making on GM food and crops has spilled over into the area of trading policy, requiring the EU to take up positions at a number of international fora, including negotiations at the WTO, the Cartagena Protocol, and the WHO (with results described in Chapter 1). Perhaps Monnet and Schuman might not be pleased at the negative nature of the exchanges between the USA and the EU and the apparent turn towards trade conflict rather than collaboration to which the arguments about GM food and crops have contributed. On the other hand, as Weale comments in general about EU environmental policy 'the functional links between the Single Market and environmental policy are not sufficient to explain the form and character that the policy has taken' (Weale 1996: 605–606).

One clear signal thrown up by this issue is that the EU is indeed a powerful actor when it acts coherently, as it has done through its regulatory policies on labelling and the action of European consumers in rejecting GM food. The USA is likely to find that its recourse to WTO litigation will fail to sell its products and defeat the practice of GM labelling. The USA faces the unfamiliar option of being forced to consider giving way to someone else, and in this case consider the option of tailoring its labelling policies to suit EU markets. Certainly the proportion of trade conducted by the EU is already more-or-less equal to that of the USA. Enlargement of the EU will further enhance its bargaining ability. I want to break down the remainder of my survey of the EU's institutional relations into three sections: Commission, member states and European Parliament.

Commission

The notion that the Commission has enhanced its own area of authority in issues where the member states and industry have been divided (Cram

1997: 176) seems true as regards the initial stages of implementation of EU regulation. Certainly the powers of the Commission were enhanced, through the 90/220 Directive, in the context of environmentalist unease over biotechnology in the late 1980s. However, as GM products came up for authorisation in the later 1990s the Commission's ability to enforce its own will diminished as the issues became ever more matters of public controversy across the various member states. Although, under the terms of the 90/220 Directive, the Commission was required to authorise proposals for GM crop and food products in the event of a qualified majority being unobtainable at the 'Article 21' Committee, it increasingly found that it was politically difficult to carry this out. I discussed earlier the problems that the Commission had in carrying CIBA/GEIGY maize into law. Although this proved to be the most controversial application, other applications received objections, and the group of 'hard line' states who were making the objections increased in number.

In April 1998 the Commission wanted to take legal proceedings against Austria and Luxembourg for their decision to ban the import of the CIBA/GEIGY (by then called 'Novartis') maize. However the Commission did not obtain a qualified majority for this action since Austria, Greece and Luxembourg objected to the Commission's proposed action and Belgium, France, Denmark and Ireland abstained (Greenpeace 1998). A qualified majority constituted 62 out of 87 Council votes, but the countries opposing or abstaining had 32 votes between them.

Matters took a turn for the worse and from October 1998 there was an effective, de facto, moratorium on new GM food and crop product authorisations. A solid block of what became six states, Austria, Denmark, France, Italy, Luxembourg and Greece announced that they would not sanction any new product approvals until new rules on traceability and labelling were brought into force. Belgium joined this group in 2001 (DG environment official interview – 2002).

In theory the Commission could have invoked existed law in the 90/220 Directive and simply overruled the objecting member states. However, as one Commission official put it 'The Commission would not take political responsibility' (DG environment official interview – 2002). The issue of countries banning products under Article 16 of the 90/220 Directive, involving member states' rights to take action if new evidence of risk becomes available, proved to be the biggest stumbling block. If bans were put in place in some states on some products, the market would be fragmented. There were nine occasions where bans have been introduced on the basis of appeals to Article 16. Austria deployed the sanction on three separate occasions, France twice, and Luxembourg, Greece, Germany and the UK all once. UK's ban came from the Welsh Assembly who demanded longer separation distances for Aventis's 'T25' herbicide tolerant maize.

We can separate out institutional influences using Scott's (1995) division between normative, cognitive and regulative rules (discussed in Chapter 2). I have already discussed 'regulative' rules, or what rational choice institutionalists would call 'incentive structures' earlier when discussing labelling and also the 'Article 16' and 'Article 21' rules giving influence to member states over GMO authorisations. We could say that a Commission 'norm' is that it will not act against a recalcitrant state if that state is acting with the support of several states. As a Commission official put it:

> If only one or two Member States are opposed to regulation then perhaps they may be drawn up before the court. If there is a big group of Member States including important Member States that are in opposition, the Commission will hesitate to take steps against them. The Commission took the other road to develop more stringent regulations to get more consensus between the Member States. It was practical to change the regulations through introducing the new food and feed and traceability and labelling regulations and also the 01/18 Directive. There would have been too big a conflict if you have nearly a majority of Member States against the Commission.
>
> (DG environment officials interview – 2003)

Otherwise the Commission wished, as far as possible, to smooth the path towards further GM food and crop authorisations. It did so for two reasons. First it wished to pursue technological development of GM food technology. This is a belief, a cognitive rule. Evidence for its existence lies in the original nature of its proposed revisions to the 90/220 Directive (CEC 1996), its action over the CIBA/GEIGY maize affair, and also the views of Commission officials as given to the author of this study. Second, the Commission wanted harmonious trade relations with the EU's principal trading partners, especially the USA. These two objectives, the former a belief (a cognitive rule), the latter a norm, conflicted with the objective of leading the EU through consensus (another norm). A further Commission norm which also conflicted with using its authority to legalise GMO product uses was the idea of maintaining the integrity of a single market. If some member states managed to ban products and the Commission did not feel it had the political authority to overcome those bans, then its only course of action in maintaining a single, unfragmented market, was to stop the process of GMO authorisations until it could achieve conditions where this fragmentation would not occur. This condition could only be achieved if the Commission acceded to the demands of the states that had declared support for the moratorium. Indeed, with the adoption by Belgium of the pro-moratorium position, the pro-moratorium states actually constituted a majority of the votes of the Council of Ministers, albeit a numerical rather than a qualified majority. Weale (1996: 606), citing Scharpf (1988: 242) points out that in systems of government where 'decision making authority is shared (such as in the EU) ... decision making must necessarily take the form of requiring concurrent majorities of actors'.

In order to get a concurrent majority (importantly, in the European Parliament as well as the Council of Ministers) it was necessary to go along with the demand of the 'pro-moratorium' bloc for the Food and Feed and Labelling and Traceability rules that were demanded by the bloc. Most other member states (with the exception of the UK government) supported, or at least were prepared to tolerate, these new rules. As if to emphasise the inability of the Commission to do anything to seriously avoid the moratorium on GMO approvals, in March 2000 the European Court of Justice actually supported a French ban on Novartis (formerly CIBA/GEIGY) maize. The French ban was instituted following a court action by Greenpeace and was upheld by the Court on the grounds that various health and environmental risks had not been taken into account when the maize was approved in the first place (Greenpeace 2000).

It would therefore be misplaced for anyone to claim that the moratorium on GMO approvals was the result of action by, or even the support of, the European Commission. Rather there are clear instances of action, such as the moratorium on new GMO authorisations, promulgated by recalcitrant member states. This action was supplemented by the European Parliament, which made a series of crucial amendments to various pieces of GMO legislation. These actions were associated with strong consumer and environmental opposition to GM food and crops. This opposition was expressed forcefully at the national level, but, as I shall discuss later, it was also transmitted to the Commission and Parliament through interest groups operating at the EU level. As I mentioned in Chapter 2, textbook theory has it that the European Parliament is a natural ally of the Commission, particularly the DG Environment. This is far from the case in GM food and crop policy.

Caroline Jackson MEP, the Chair of the European Parliament's Environment Committee, speaking in June 2003, just before the second reading of the Food and Feed and Traceability legislation, commented:

> I think the chemistry is very strange because (while) I think that the Commission wants to unblock the moratorium, I think that the majority in the Council probably doesn't want to unblock the moratorium. The Commission wants to get the (food and feed and labelling and traceability) legislation through in the knowledge that if it fails, each member state may go its own way and fragment the market. So I don't think this is a usual way of seeing the Parliament and the Commission on the same side. In many ways, I think that many members of the Parliament aren't on the same side as the Commission because they believe that the Commission's not gone far enough and the Commission doesn't want to be pushed to go any further. Very often you find

that the Parliament wants to push the Commission and the Commission wants to be pushed. In this instance I think the Commission doesn't want to be pushed because it realises that it is likely to find that the (US's) WTO case is thereby strengthened and that some member states, notably the UK, will say this is absurd.

(Jackson interview – 2003)

Please forgive me for the length of the above quote, but it does sum up some relevant points very well, and is significant as a statement in its own right since the speaker is a leading actor in this issue. Having focused on the motivations and actions of the Commission, I now want to have a look at the positions of the member states.

Member states and the Council of Ministers

Given the size of this book, it is not possible to conduct a full survey of GM food and crop politics in all member states, although Chapter 3 does represent an attempt to focus on the details of the debate in one state, the UK. It is impossible, anyway, to disentangle the influence of the various national political cultures on the Council of Ministers and on the Parliament. What, perhaps, is possible, is to say something about the formal role of the governments on the GM food/crops issue and compare this with the results of public opinion surveys.

It is impossible, for philosophical reasons which I described in Chapter 2, to ascribe specific influence to agents which is separable from context. However, we can talk about the actions that agents have taken. As may be gleaned from the preceding analysis, the 'pro-moratorium' states were instrumental, in formal terms, in forcing the Commission to introduce new legislation to strengthen assessment of GM crops (through Directive 18/01) and in widening the scope of labelling through the Food and Feed and Labelling and Traceability Regulations. The pressure of the GM-sceptical governments on the Commission has also contributed to the achievement of a compromise on 'coexistence' issues, although this change was formally achieved through an amendment at the European Parliament. What I want to do in this section, is to look at the relative positions of the member states, say something about the timing of their relative concerns and also see what sort of association there is between the positions of the governments and their respective public opinions as measured through the *Eurobarometer* opinion survey.

As can be seen from Table 5.1, scepticism of GM food is significant in all countries, and support for GM food is very low compared to the support for medical biotechnology applications, as represented here by genetic testing. Even in Austria, where the support for genetic testing is the lowest in the EU, this level of support is still higher than the support for GM food in any country. It is also interesting to note that support for

Table 5.1 National changes in support for applications of biotechnology 1996–2002

	Genetic testing			GM crops			GM food		
	1996	*1999*	*2002*	*1996*	*1999*	*2002*	*1996*	*1999*	*2002*
Austria	74	78	78	39	41	57	31	30	47
Belgium	95	90	92	89	74	80	72	47	56
Denmark	91	91	93	68	58	73	43	35	45
Finland	95	91	94	88	81	84	77	69	70
France	96	94	92	79	54	55	54	35	30
Germany	87	90	85	73	69	67	56	49	48
Greece	97	91	92	77	45	54	49	19	24
Ireland	96	94	94	84	67	77	73	56	70
Italy	97	95	95	86	78	68	61	49	40
Luxembourg	91	85	91	70	42	54	56	30	35
The Netherlands	93	96	96	87	82	85	78	75	65
Portugal	97	96	93	90	81	84	72	55	68
Spain	96	94	94	86	87	91	80	70	74
Sweden	92	92	93	73	61	73	42	41	58
UK	97	96	95	85	63	75	67	47	63

Source: adapted from Gaskell *et al.* 2003, Table 5 p. 18.

GM crops is higher than for food. Perhaps this is a reflection of consumer concern with food health issues being stronger than their concern for, or awareness of, environmental issues concerned with crops. Arguably, the way Gaskell *et al.* set out this table gives a rose-tinted view of GM techno-logy since they include 'risk tolerant' respondents, that is those who still harbour suspicions of the technology, as 'supporters'. However, this is sec-ondary to the comparative function of the table, both in regard to differ-ent applications of biotechnology and also in regard to other member states.

Those who remember the earlier coverage of this chapter will notice that there seems to be a high correspondence between member states who expressed various types of opposition or precaution to GM food and crop authorisations, and those member states which record low levels of support for GM food. I mentioned earlier that Austria, Denmark and Sweden objected to Monsanto's GM soya being authorised in early March 1996. These are the very same countries that, according to Table 5.1, had the lowest levels of support for GM food in 1996. Another association may be even more significant. Seven out of the eight states with the lowest support for GM food in 2002, in the table, supported the moratorium on GM food and crop authorisations.

We can see some interesting associations of specific policy stances taken by member states and the timing of those stances. For example Sweden was an early opponent of GM food and crop authorisations, but later on the Swedish government has taken up a less negative stance. The UK government's negotiation of a voluntary moratorium on cultivation of GM crops in 1999 coincided with low support for GM food. France and Italy both joined the GM-sceptical group of nations later than Denmark and Austria who record low levels of support for GM crops as early as in 1996.

It would, for philosophical reasons which I have discussed earlier, be unwise to make a necessary causal link that says that governments take stances towards GM food simply because of the nature and strength of public attitudes towards the technology. However, it does seem less plausible to argue that governments lead public opinion rather than the other way around. This is because of two arguments. First, if government were leading public opinion rather than the other way around one would expect to see some pro-moratorium states, or some states which had objected to specific GM food/crop authorisations, which did not have the support of the public. In fact the opposite seems to be the case. One can see that the German government, for example, has resisted a strong current of public opinion that is opposed to GM food, despite the fact that it includes the pro-moratorium Greens (Die Grunen) and a Green Environment Minister.

The only countries that have supported the moratorium on GM food and crop authorisations have been countries where there was a strong current of public opposition to GM food. Besides this, it seems intuitively odd for governments to ignore the advice of their civil servants and industrial interest groups in favour of radical environmental groups without some other incentive coming into play. In the case of France, for example, the Conservative Juppé government seemed very much in favour of GM crop technology until blown off course by an upsurge in concern at the beginning of 1997. Certainly, France used to be the favoured place for GM crop trials. However, the Juppé government decided to ban commercialisation of CIBA/GEIBY maize in February 1997, an act which prompted the resignation of Axel Kahn, the Director of the Biomolecular Genetics Commission which was responsible for GM crop regulation (Reuters 1997).

I mentioned in Chapter 4 how the organic movement has provided much of the activist support for the opposition to GM food and crops in the USA. There is an association between the support for organic food in the EU and opposition to GM food. According to a survey compiled by the UK's Soil Association (Soil Association 2001) Denmark and Austria have the highest sales per head of organic food produce in the EU. However, in sociological terms, broader associations can be made between those countries that have had strong, mass, activist movements against nuclear power in the 1970s and 1980s. Both Austria and Denmark had very active anti-nuclear movements in the late 1970s and early 1980s which were successful in defeating government proposals to embark on a nuclear power construction programmes. The German anti-nuclear movement has also been very militant. The Danish and German gene technology legislation that was discussed earlier was adopted following pressure by a Green movement whose critique of the alleged 'abuse' of genetic technology by undemocratic leading corporations for dangerous ends, had much in common with the anti-nuclear movement's critique of the role of nuclear technology. The work of Habermas (1971) emerged in the same environment of suspicion of the 'positivist' use of science that benefited corporate interests. Beck's 'Risk Society' thesis, first published in 1986, highlighted both nuclear power and genetic engineering as technologies which posed potentially uncontrollable risks (Beck 1998). In an analysis of biotechnology regulation in Austria, Mikl and Torgersen (1996: 195) comment that:

> The disapproval (of gene technology) was comparable with that in Germany ... One reason (for the disapproval) may be that public trust in expert opinion declined during the 1980s, after Austrian voters had rejected nuclear power in a 1978 referendum.

As can be seen in Table 5.1, Spain appears to be the most favourably disposed towards GM food. It is the only country in the EU where any GM crops are grown on a commercial (as opposed to trial) basis. Around 6 per cent of the 2002 maize crop was Bt maize. However, as one news report put it, 'Public opposition in Spain is enough to make farmers guarantee they are selling GMO-free to maize millers, who produce starch for human consumption' (Roberts 2002).

The UK government is regarded as the biggest friend of agricultural biotechnology in the EU. The UK government has distanced itself from the Food and Feed and Traceability and Labelling Regulations, just as in the case of the BST (beef hormone) dispute between the USA and the UK. On the other hand, as is the case with the rest of the EU, no GM food is sold in UK supermarkets. Moreover the government negotiated a moratorium on GM crop commercialisation while the FSE Trials (see Chapter 3) were being conducted. One has to comment that if the UK and Spain are the biggest friends of GM food and crops in the EU, then the positions of agricultural biotechnology in the EU is, indeed, parlous! The results of the UK's FSE trials will only further encourage those states that have objected to authorising GM crops to continue their blocking actions.

The Eurobarometer survey does suggest a small shift in favour of EU support for GM food in the 1999–2002 period. Nevertheless, it does seem that opinion in the various EU member states will have to shift a very long way before there is any prospect of significant GM food sales in the EU. Biotechnology companies are hoping that the enlargement of the EU will shift the EU towards their interests. How realistic is this?

EU enlargement

Ten Central and Eastern European states become formal EU member states in April 2004. How might this affect the position of GM food and crops in the enlarged EU? Certainly biotechnology companies hope that the political influence of member states who supported the moratorium on GM authorisations will be diluted. Commission officials think that there may be more opportunities for doing GM crop trials in the new member states.

However, these factors may be more than counterbalanced by other, more negative, factors. First, the GM food labelling regime, including the labelling of derivatives and animal feed will be applied in these new states, which will do nothing for GM food sales. Second, support for GM food and crops in Eastern Europe is not as solid as some may believe.

Farmers in parts of Eastern Europe have developed an interest in supplying EU markets with non-GM supplies that substitute for supplies of crops from places, mainly North America, which produce GM food and who do not segregate that production from non-GM food. Polish farmers now supply oil seed rape which used to be derived from Canada. Polish farmers are keen to keep hold of their gains in supplying oil seed rape as a substitute for supplies that were previously brought in from Canada. Indeed, the Polish Consumers Union has adopted a very strict position which favours 'zero contamination' (of conventional or organic crops by GM crops), which goes beyond that which is being demanded by the BEUC, the European Consumers' Organisation.

Indeed, the Consumers' Organisations in the new member states appear to be generally in line with the other, Western European, affiliates to the BEUC (Kettlitz interview – 2003). It could well be that the enlargement of the EU will actually significantly reduce the market for GM food technology. At the end of the day consumers in Eastern Europe see no significant advantage for GM food and their farmers do not wish to do anything that could jeopardise markets in existing EU countries. It seems very unlikely that the new member states will orchestrate a drive for EU legislation that is more friendly to the interests of agricultural biotechnology than that which exists on the statute books at present.

European Parliament

As one analyst, describing the impact of the Environment Committee of the European Parliament puts it, 'The practical legislative impact of the Committee stands in stark contrast to the academic assessment which still maintains that the European Parliament itself is not very powerful, lacking true legislative capabilities' (Judge 2002: 137 citing Thomas 1992: 4–5).

The Parliament emerges as a major influence on GM food and crop regulation policy to such an extent that it is difficult to deny the notion that, on this issue, it has been a partner of equal importance to both the Commission and the Council of Ministers. Even when the seemingly radical amendments proposed by the Parliament have not been adopted, they have, more often than not, been enshrined in later legislative proposals generated by the Commission.

We can begin this brief survey of key Parliamentary examples of GM legislative influence with the 90/220 Directive. Lake (1991) has made an analysis of the effects of deliberation by the European Parliament on the Directive. Among various changes, the Parliament (upon the recommendation of the Environment Committee) persuaded the Commission to separate out the 'contained use' (90/219) Directive from the 'deliberate release' (90/220) Directive. The Parliament passed amendments requiring that the issues of legal liability and compensation were dealt with. This was not accepted by the Commission. The Parliament also passed amendments requiring member states 'to ensure that adequate information is provided in advance of a planned release to the inhabitants of the areas concerned' (Official Journal of the European Communities 1990b, cited by Lake 1991: 12). This was removed from the final draft by the Commission. However, as discussed earlier, consultation was made mandatory in the 18/01 Deliberate Release Directive. The Parliament also strengthened the requirements to rule out adverse effects on health and the environment in the 90/220 Directive.

As I described earlier, the European Parliament was central to the requirement that GM food be labelled. This development came about as a consequence of Parliamentary amendments to the Novel Foods and Novel Foods Ingredients Regulation Number 258/97. Parliamentary pressure (in the form of the censure resolution) also contributed to the Commission proposal for a regulation covering GM maize and soya (1139/98) which had been authorised before 258/97 came into force.

When Directive 18/01 came before the Parliament it was subjected to a range of amendments presenting further restrictions and checks on GMO authorisations. Once again the Parliament passed an amendment requiring liability rules to be introduced to cover cases of damage to the environment, although this was rejected by the Commission. The Parliament was successful in excluding genetically modified human beings from being subject to the Directive, but an amendment from the Environment Committee including 'DNA and plasmid' (Official Journal of the European Communities 1999: 366) in the definition of a GMO was unsuccessful.

The Parliament voted to ban all uses of antibiotic resistant marker genes, a formulation which the Commission rejected. However, in the final legislation a compromise formula involving phasing out use of medically important antibiotic marker resistant genes was agreed (as discussed earlier). The Parliament was successful in ensuring mandatory consultation on proposals for authorisations. It was also successful in gaining a commitment from the Commission to introduce legislation approving the

Cartagena Protocol and also a commitment to bring forward legislation on labelling and traceability of GM food. The Parliament also passed an amendment giving it the power to 'act by an absolute majority of its members' to turn down an application for GMO release authorisation (Official Journal of the European Communities 1999: 377), but this was unacceptable to the Commission and Council.

The Parliament's amendments to the (later) Food and Feed and Labelling and Traceability Regulations 1829/03 and 1830/03 provoked a major controversy over so-called 'thresholds' for declaring food products to be GM-free. The Commission proposed a maximum threshold of 1 per cent of 'contamination' by GM food. In the first reading the Parliament passed an amendment lowering this threshold to 0.5 per cent and set down the notion of 'zero tolerance' for unauthorised GMOs. The Commission, arguing that this was unenforceable presented a compromise, which was accepted at the second reading. The compromise involved the threshold being set at 0.9 per cent. Furthermore unauthorised GMOs could account for up to 0.5 per cent of the content provided that the GMO in question had gone through some sort of scientific assessment. Amendments, sponsored by the Green Group of MEPs, were approved requiring GMO producers to take anti-contamination measures. As mentioned earlier, a compromise was reached at the second reading regarding this 'coexistence' issue which gave discretion to member states to take action necessary to avoid contamination.

Despite the sometimes influential intervention of the Green Group of MEPs, the main divisions on votes within the European Parliament have been formed around national position rather than positions held by party groups (Jackson interview – 2003; Whitehead interview – 2003). It does seem that there is a close association between the votes of MEPs and the strength of public concern on GM food in their respective countries. This underlines the lesson that when we discuss the 'influence' of the member states, we are not just talking about their formal influence through the Council of Ministers, we are really talking about the influence of national political cultures as channelled through the Council of Ministers and the European Parliament. Caroline Jackson, the Chair of the Environment Committee described some key influences on her in the GMO debates as:

> We can certainly tell the (UK) national mood perhaps from the press, but the press are very alarmist on this issue. And then I use as a counter balance Krebbs and May, the FSA (Food Standards Agency) and the Royal Society, respectively.
>
> (Jackson interview – 2003)

We can see in this quote an example of the sort of interaction between public, scientific and political attitudes that is a central aspect of the 'discourse-paradigm' approach. Such interactions lead to the development of distinctive national regulatory paradigms on the GM food and crop issues. Such nationally based discourse-paradigms could be said to exert a composite effect through the agency of the European Parliament. The Parliament contains representatives which have been immersed in the differing discourse-paradigms that are dominant in the various member states. It would, however, be wrong to assume that all interest groups relevant to the GM food and crops issue are represented solely at the national level. All the main interests are well represented through 'peak' organisations which are constituted at a pan-European level. As has been the practice in the chapters on the UK and the USA, I shall deploy 'network' analysis in order to analyse these pan-European interest groups.

Pan-European networks

The usefulness of using network analysis based on the adoption of common storylines rather than use of an advocacy coalition approaches based on the possession of common core interests is aptly demonstrated at the EU level of this case study. The 'discourse-coalitions' which I analyse consist of actors whose central interests appear to be quite diverse. Although there is a tendency for the producer-interest groups to be more pro-GM food and crop than 'cause' or 'consumer' groups, the discourse coalitions in fact are mostly made up of mixtures of producer and cause or consumer groups.

The pattern of influence of the various interest groups vis-à-vis the European Commission can only be described, using the Marsh and Rhodes (1992) typology, as an issue network rather than anything even remotely resembling a policy community. In order to understand the pattern of positions on the issue it is useful, in some ways, to see it from the point of distance from the point of production. On the one side there are the consumers, many of them hostile to GM food and being actively organised to boycott GM food by environmental groups. Consumer opinions are articulated by consumer organisations. Then, as a spokesperson for the European Food and Drink Organisation (CIAA) put it:

> Those consumer organisations impacted on the retailers, saying 'we don't want those products'. So why should the retailer then offer that product? Then the retailers tell the manufacturers 'we don't buy those products any more. We won't sell them'. What can you do? So the manufacturers say to their suppliers 'We want non-GM ingredients, non-GM raw materials' and then you get the farmers.
>
> (Taeymans interview – 2003)

I find it convenient to divide up the networks associated with the various discourse coalitions into three basic networks: the food production network, the retail network and the anti-GM network.

The food production network

The common storyline here is that GM food technology is a legitimate part of the food industry, and while the acceptance of labelling of GM food is necessary, the process of regulation has been pushed to extremes by unfounded food scares. Key actors supporting this type of discourse are the food manufacturers and the biotechnology companies. Mainstream farmer organisations tended to be supporters of this storyline in earlier years, but they have become more disconnected from the food production network in more recent times.

The organisation with the most regular contacts on the GM food issue among the food producers is the CIAA, the European Food and Drinks Federation. They are parallel with the Grocery Manufacturers of America and represent the big manufacturers like Danone, Kraft or even Coca Cola, which has plants in Europe. The CIAA also networks with organisations like Europen, which represents the European packaging industry and UNICE, the Union of Industrial and Employers' Confederation of Europe. The pan-European organisation representing biotechnology is EuropaBio.

Both EuropaBio and the CIAA originally opposed the amendment to the 1997 Novel Food and Novel Food Ingredients Regulation which mandated the labelling of GM food on the basis of detectability of DNA, although they later came to accept it as a demand that was needed to satisfy consumer expectations. However, they have been strong opponents of the moratorium on GM crop and food authorisations. They have also opposed the notion of labelling on the basis of 'traceability' (as opposed to DNA detectability). The traceability notion has been enshrined in the recently approved regulations.

Regardless of such policy concerns, however, the CIAA affiliates have had no choice but to remove GM food from their products. They will have to follow suit by finding substitutes for processed foods that are derived from GM food products as well if, as seems likely, this is demanded by the retailers.

Although the farmers are nominally supporters of what could be described as the 'food producers' storyline on GM food and crops, the fact that practically no European farmers grow GM crops for commercial purposes has allowed them to take a more dispassionate stance than that of US farmers. However, it is also true to say that the mainstream farmers organisations want the opportunity to use GM food technology and they have a track record of opposing the trend towards imposing greater restrictions on GM food and crops.

European farmers are represented at the EU level by COPA/COGECA. They are keen to stress that the consumer must be reassured about safety and labelling requirements. However, they describe the system of labelling GM foods by traceability as being impractical (COPA/COGECA 2002).

Radical Greens and their allies have been demanding low threshold requirements for GM food labelling, and also big crop separation distances to reduce the chances of 'contamination' of conventional and GM crops. By contrast, COPA/COGEMA state that 'COPA and COGECA cannot accept the fixing of thresholds or unrealistic parameters relating to the distance to be kept between GMO and non-GMO crops, or to the time delay between the planting of two crops (previous cropping)' (COPA/COGECA 2002). There are farmer groups, chiefly representing small farmers, who have vigorously opposed GM crops. However, even in France where anti-GM farmers have, perhaps, been most visible, the main farmers organisation, the FNSEA, was the loudest to protest when the French government began to turn against GM crops in early 1997 (Reuters 1997). COPA/COGECA do side with critics of the GM crop industry in the demand that conventional and organic farmers should not be liable for damage resulting from use of GM crops.

The consumer-retailer network

This network, consisting of the biggest consumer and retail pressure groups, has focused on measures to improve the provision of labelling information on GM food products and other additives for consumers. A key discourse has been 'freedom of choice for consumers' with the implicit idea that consumers be given the opportunity to choose non-GM food. Indeed the retailers emphasised that they were in favour of GM food, but wanted labelling to build consumer confidence. The storyline was not anti-GM, but later on this became contradictory as retailers simply took GM food off the shelves. Nowadays European consumers will, usually, have to turn to the USA if they want to buy GM food! The two key organisations in this network are EuroCommerce, which represents the main supermarket chains, and the BEUC, the European Consumers' Organisation. BEUC has the principal national consumer organisations as its affiliates.

The early conversion of the BEUC and its affiliates to the notion of labelling of GM food was a key milestone in the European debate on GM food, and this was associated with the adoption of labelling in the Novel Food and Food Ingredients Regulation (258/97). Where the Consumers Unions have led, the retailers have tended to follow. I think it is significant that the Consumer Unions were simultaneously pressing for additives and flavourings to be labelled at the same time as GM food. This was also incorporated into regulation 258/97. The point is that there is European consumer pressure for information on all ingredients which are not seen to be 'natural'. This is a key distinction to be made with many US consumers.

Although in 1996 most attention was on the principle of labelling GM food on the basis of DNA detectability, the focus moved onto 'traceability'.

This occurred in response to demands from consumer organisations (and also anti-GM groups) that processed foods be labelled as well. By the summer of 1998 even the UK-based Sainsbury's supermarket group was 'reviewing' its policy on derivatives having been put under pressure by the UK Consumer Association. The Director of the Consumer Association said 'Many people don't want this (GM food) technology for a variety of reasons. The current position by government and suppliers denies the consumer choice' (Vidal 1998b). At the EU level, EuroCommerce allied with the BEUC in pressing for the labelling of food products derived from GMOs.

The Commission responded by including labelling of such derived products in its proposals for Regulations on Food and Feed and Labelling and Traceability. Xavier Durieu, the Secretary General of EuroCommerce said that 'It is through the combination of strict traceability and accurate labelling that we will give consumer informed choice and confidence in the use of GM food technology in food products' (EuroCommerce press release 2002). However, the retailers opposed radical green attempts to extend labelling to the use of GM enzymes and animals reared on GMOs. The BEUC did not promote these radical amendments. Neither did it promote amendments, passed at the first European Parliamentary reading of the Food and Feed and Traceability Regulations, to reduce thresholds beyond that which was thought practical by the Commission (Kettlitz interview – 2003).

The anti-GM network

What is really an 'anti-GM' network can be associated with a storyline that demanded that the EU's moratorium on new GM food and crop authorisations continued until effective rules are put in place which:

1   guarantee food and environmental safety;
2   ensure full transparency through labelling of all uses of GMOs in food processes;
3   ensure, in the 'coexistence' measures that all possible 'contamination' of conventional and organic crops be avoided.

Many would say that such a strategy amounts to banning GM food and crops by the attrition of numerous restrictive rules. Supporters of this 'pro-moratorium' and 'anti-contamination' discourse include a diverse collection of interests. Key ones are: Friends of the Earth, Greenpeace, the Federation of Green MEPs, some of the more radically oriented national consumer organisations, the organic farmers organisations, some groups representing small farmers and EuroCoop which is the cooperative retail movement.

Although representing a relatively small proportion of European farmers, the Confederation Paysanne Europeenne (which is an extension of José Bové's French Confederation Paysanne at a European level) has been a particularly bitter opponent of GMOs. It has even engaged in direct action against GMO crop trials. EuroCoop, which represents the European Cooperative Retail Movement is a hybrid consumer/retail organisation. Its affiliates claim a combined membership total of over 20 million. EuroCoop has worked closely with the anti-GM groups. It supported the objectives of not lifting the moratorium on GMO authorisations until regulations on labelling and traceability were in place. It called for anti-contamination measures 'all along the food chain' in order to protect consumer choice (EuroCoop 2003).

In the 1996 to 2002 period the anti-GM activists focused on support for the moratorium, but since the introduction of the labelling and traceability proposals their emphasis has shifted towards making sure there is no 'contamination' by GM crops. Hence their support for very strict, low, thresholds of 'contamination' by GMOs, over which seeds or food should be labelled as containing GMOs. This position was passed by the European Parliament at the first reading of the proposed Food and Feed and Labelling and Traceability Regulations. I have (earlier) described the details of the eventual compromise reached with the Commission.

The anti-GM network just failed to obtain a simple majority in the Parliament for its demand that meat from animals reared on GM animal feed be labelled. Nevertheless the Greens appeared to be quite satisfied with the final form of the regulations, especially the concession on 'coexistence' of GM and non-GM farming. Indeed Greenpeace published a press release welcoming the regulations with a title 'European Parliament Paves the Way for GMO-Free Europe' (Greenpeace 2003). However, Greenpeace lamented three alleged shortcomings of the legislation: that the regulations did not mandate the labelling of meat and dairy produce reared on GM feed; that 'coexistence' measures are not mandatory on all states, and that seed producers may be allowed a 'contamination threshold' of up to 0.7 per cent under which seeds need not be labelled as containing GMOs if the proportion of GM seeds is less than 0.7 per cent.

Conclusion

A key conclusion of this chapter is that we must discount the notion that the BSE crisis was the principal cause of the regulatory problems faced by GM food technology in the EU. By the time of the BSE crisis, a number of important developments had already occurred that clearly distinguished the EU from the USA in its attitude towards GM food technology. There were quite different regulatory structures (compared to the USA) in place in the EU by way of Directive 90/220. The EU had banned rBST growth hormone used in cattle. Member states such as Austria, Denmark and Luxembourg were opposing GM food and crop authorisations before the BSE crisis erupted. Moreover, moves to label GM food were already well-advanced before 20 March 1996 when the British government announced that 'new variant' CJD had been found in humans.

Often, explanations for the EU's hostility to GM food and crops refer to European regulatory failure to deal with food crises. However, it is rather sloppy thinking, for example, to start talking about the dioxin-in-food crisis in Belgium being connected with hostility with GM food. By the summer of 1999, when this crisis took place, there was already an EU moratorium on GM product authorisations in place and GM food had mostly disappeared from European supermarket shelves.

Now it is plausible to argue that the BSE crisis reinforced an existing discourse of mistrust of food quality and of regulators, and this may especially be the case in the UK. However, even here it is very difficult to untangle association from cause. It is also hard to distinguish between the BSE crisis being significant mainly for providing a new metaphor used in rhetoric against GM food and, alternatively, it being an actual cause of changes in attitudes to food health risks and food quality issues.

This case reveals the problems and shortcomings of approaches such as the advocacy coalition framework which rely heavily on the notions that changing events produce outcomes. The events have to be interpreted, and once advocacy coalition framework advocates concede that intepretation of events matter, then any attempt to anchor analysis principally on specific events becomes unstuck in the shifting sands of context. Moreover, there is often a danger (which is illustrated in this case) that sloppy post-hoc analyses may slide over important evidence to produce a simplified picture of the world that appears to conform to cause-and-effect modelling.

Likewise, this case study should illustrate the shortcomings of analysis based mostly on analysis of agents. We should be sceptial of 'blaming' particular agents for outcomes. Americans may blame the EU's hostility towards GM food on the aggressiveness of environmental groups and the timidity of European food retailers and producers. Yet such agents operate within an EU cultural context that means that agents' actions will have very different connotations and consequences compared to actions of environmentalists or retailers in the USA.

This case study also reveals the extent to which scientific truths are culturally bound. According to GM-sceptical discourse, for example over the CIBA/GEIGY maize case, it was European politicians who were abandoning scientific precaution in favour of political/economic trade expediencies. The EU has developed a different set of scientific assessment procedures, and has often emphasised a different set of tests for environmental acceptability compared to the USA. The US authorities, by contrast, often assert that the EU is ignoring science in favour of politics.

This difference in scientific judgement does not mean that either the USA or the EU have necessarily got it 'wrong', merely that they are making different judgements about uncertainty, and also different judgements about what is important in an environmental sense. For example, the European system places emphasis on problems of weeds acquiring resistance to herbicides and on tests on the effects of GM crops on food supply for wildlife. It is just that different 'discourse-paradigms' place emphasis on different values and standards for assessment. The different scientific judgements relate to different cultural circumstances which interact with the various scientific communities in the member states. The influence of the European Parliament is based on scientific/cultural judgements drawn from a composite of different national 'discourse-paradigms'. The European Parliament gives a voice to the political demands that emerge from these 'discourse-paradigms'.

This case certainly acts as a counterveiling example to textbook theories that the Commission supports the pro-environmentalist leanings of the European Parliament and its Environment Committee against the member states. The Commission, in this case, has been on the defensive and has tried to soften public demands for restrictions on GM food and crops. Perhaps an explanation for this is that the Commission feels that it ought to pursue its normative role of producing harmonious trade relations, especially in pursuit of what the Commission still regard as an area of technical progress. As regards other questions of EU theory raised in Chapter 2 we can say: that in this case there is much evidence of functionalist spillover of harmonization of rules into international relation issues, albeit by accident rather than Commission design; Majone's (1992) notion that industrialists press for environmental regulation only applies to a wish for minimalist approach to such regulation; there is great evidence of effective influence on legislative outcomes by the European Parliament; on the other hand, in this example, the Commission (contra Cram) has spectacularly failed to impose its will on recalcitrant member states. It would be wrong to try to rewrite the EU theory textbooks completely on the basis of this case study. Rather it would be better to say that there is evidence, in this case, that the textbook theory applies better to cases where there is not the type of major public controversy that applies to the issue of GM food and crop politics.

US observers may have become accustomed to framing their understanding of agricultural politics in the EU by witnessing a battle between the Commission and farming interests, the latter being determined to defend the productionist subsidies of the Common Agricultural Policy. As a result these US observers might be lulled into feeling that it is farmers, and not really consumers, who are the ones who are rejecting GM crops, and doing it for protectionist reasons. However, this seems to be a mistaken view. Despite the headline grabbing activities of José Bové, mainstream farmers have, on balance, been keen (at least in the initial stages) to be offered opportunities to try out GM crops. They have certainly tried to persuade the EU not to adopt the more restrictive measures on labelling, thresholds and 'coexistence' which have been proposed by radical environmental and consumer groups and often backed by the European Parliament.

I have attempted in this chapter to contrast use of rational choice style focus on incentive structures, or regulative rules as Scott (1995) calls them, with discourse analysis. I do not deny that sometimes such rational choice approaches can have uses, although I contend their use is more often concerned with the provision of analytical framework than an ability to produce useful predictions. But two points emerge. First, that use of Foucault's notion of power and knowledge as deployed through surveillance means much the same thing as an 'incentive structure'. However, in Foucauldian analysis, this power/knowledge is based on a particular dominant discourse. Second that the key force through which incentive structures are constituted and have their effects is that of the belief system, or cognitive structure. These discourses can be studied through discourse analysis and the analysis of the contexts in which they proliferate and in which key agents are immersed.

6 Deliberation and GM food politics

We have seen from the theory in Chapter 2 that the public's notions of risk are different from those of the experts, whom they do not always trust. Given this gap, it seems especially pertinent to consider ways of involving the public in decisions about risk. This chapter consists of two sections. The first discusses the theory of public involvement. The second section looks at some practice of public involvement in the GM food and crop issue in Europe and the USA.

Arguments for public involvement

If one takes the position that the views of consumers who reject GM food are based solely on irrational ignorance of science which deserves not to be taken into account, one does not, logically, seek to involve consumers in decision making. That is, except to the extent that they can be induced to 'see the light' as shone forth by the expert committees who have produced the 'legitimate' verdict on the food and crops. This will involve publicity campaigns and other mechanisms to educate the public. Irwin calls such an approach a 'science-centred' perspective:

> The view is that the public forms a barrier to intelligent and constructive debate ... These representations of the public became most visible in the BSE case – with numerous references to 'public hysteria' and 'media hype'. For the 'science-centred' view, continued public concern after scientifically-based reassurances had been given could only be the product of an emotional and badly informed public.
>
> (Irwin 1995: 27)

As we saw in Chapter 3, the great pitfall of the 'science-centred' perspective is manifested all too graphically when the words of official scientists are later publicly retracted. This happened spectacularly in the BSE case when, contrary to previous official assurances, the UK government admitted that the disease had crossed to humans. This sort of result persuades consumers that their guess about potential risk is as good as the 'official' scientists.

There is a powerful argument that the lay public should be more directly involved in making decisions about controversial environmental and related issues. One can argue for the involvement of the lay-public in such decisions on at least two grounds. First, on a rights basis, in that controversial science-policy issues are concerned with making decisions about values, and not merely about technical issues, and that all citizens should have a right to be involved in the debates about values. The second argument is more practical and also utilitarian in that it might maximise benefits to society. There is no point in governments taking decisions that will be actively opposed and thwarted by a majority of citizenry when this dislocation could have been avoided by giving them real involvement in to the decision making process in the first place.

There has been an increasing literature on what has been called, notably by Lee (1993), 'civic science'. At a general level O'Riordan says that civic science entails:

> A form of science that is deliberative, inclusive, participatory and designed to minimize losers. Its purpose is to recognize that groups in society have to be involved if fairer and more comprehensive decisions are to be made. It also accepts that certain types of scientific uncertainty cannot be handled by traditional peer review procedures.
>
> (O'Riordan 2000: 9)

Lee's own conclusions on civic science were made after extensive experience with, and study of, the North West Power Planning Council which oversees ecological management in the Colorado Basin in northwestern USA. Hydro-electric dams are the centre of attention. The issues involved include wildlife protection, energy policy and industrial demands on resources. Wildlife conservation, in this case, covers a variety of pressures from different interest groups whether they be from indigenous peoples protecting fishing resources, recreational fishing groups, commercial fisheries or from groups concerned with the preservation of flora and fauna and habitats for purely ecocentric purposes. This case study provided Lee with an excellent backdrop to his ideas of how political processes and scientific expertise can interact to optimise political, social and economic outcomes. His study reveals, in the case of the North West Power Planning Council, how apparently diverse interest groups can negotiate arrangements to achieve common, negotiated objectives and how experts can help this process.

Lee uses and adapts some philosophy of John Dewey's in an attempt to shape some guidelines for the most useful approach to civic science. Dewey argued that the political process could set up hypotheses which could be tested by experts who would act as 'teachers and interpreters' to 'decipher the technological world for citizens and enable them to make sensible political judgements' (Lee 1993: 92).

Lee's expectations of scientists are lower than those of Dewey in that Lee sees the role of scientists as being merely interpreters rather than also being teachers as suggested by Dewey. However Lee has faith in the ability of a corpus of experts to interpret the negotiated aspirations of a coalition of interests in a way that involves adaptation in the face of the results of experimentation and changing conditions. He called this a process of 'adaptive management'. Lee's work emphasises the notion, which was discussed earlier in this chapter, that expertise can only produce meaningful positivistic 'truths' in the context of an agreed set of values and belief systems.

The problem, though, with this 'civic science' approach is that it does not work very well when the perceived interests of different groups are irreconcilable, as seems to be the case with some interests in the case of GM food and crops. Before we can have the sort of 'adaptive management' about which Lee so eloquently writes we have to have a firm basis of agreement on the values, the belief systems, which form the basis of any strategies which may benefit from such 'adaptive management' techniques. In short, we need what is known as deliberative democracy that involves the lay citizens.

Deliberative theory

It is difficult to imagine the involvement of lay-publics in science-policy decisions in any way that does not involve deliberation, that is the reflection on preferences and exposure to conflicting views. Let us look at some theory underpinning deliberative democracy. Having done this, we shall be in a better position to analyse the extent of deliberative democracy occurring in practice on the GM food and crops case and we may also be in a better position to make some suggestions for how this process might be improved. Dryzek (2000), an environmental scholar by training, in what is widely seen as a seminal discussion of deliberation describes the concept in the following way:

> Deliberation as a social process is distinguished from other kinds of communication in that deliberators are amenable to changing their judgments, preferences and views during the course of their interactions, which involve persuasion rather than coercion, manipulation or deception. The essence of democracy itself is now widely taken to be deliberation, as opposed to voting, interest aggregation, constitutional rights, or even self-government.
>
> (Dryzek 2000: 1)

In particular Dryzek wants to stop what he sees as 'the assimilation of (deliberative democracy) to liberal constitutionalism' (Dryzek 2000: 175). Habermas is seen as being a prime mover in this process given that, at least according to Dryzek, Habermas's later work emphasises the importance of elections as a focus for deliberation. Dryzek calls the emphasis on elections 'old fashioned' and asks; 'What are we to make of the multiple channels of influence that for better or worse do not involve elections – such as protests, demonstrations, boycotts, information campaigns, media events, lobbying, financial inducements, economic threats, and so forth?' (Dryzek 2000: 26) Nevertheless Habermas is still firmly on record in his later work as advocating a wider role for deliberation. He says:

> Deliberative politics thus lives off the interplay between the democratically institutionalized will formation and opinion-formation. It cannot rely solely on the channels of procedurally regulated deliberation and decision making.
>
> (Habermas 1996: 308)

Hence Dryzek is not so much arguing directly against the mainstream trends, as represented by Habermas and others, as trying to switch the emphasis. Dryzek comments that 'My argument is not that dispersed control over the contestation of discourses and associated networks should completely replace the more familiar sources, but that the world will be a greener and more democratic place to the extent their relative weight increases' (Dryzek 2000: 160). Certainly Dryzek argues effectively against liberals and social choice theorists. Their assumption that democracy involves little more than an appropriate way of aggregating existing preferences clearly biases politics against those who wish to alter value systems (and therefore preferences) through argument and protest. Dryzek confronts the arguments of social choice theorists that democracy can be manipulated by organised minorities. He highlights the importance of deliberation as a means of persuading citizens and deliberators to set terms of debate that will limit manipuation of democratic processes.

Dryzek affects to distance himself from what he calls 'difference' democrats such as Mouffe (1995). These 'difference democrats' attack Habermas's 'ideal speech' situation for imagining a much greater degree of common sets of values, that is identity, that can really exist in a given society. Mouffe, who is influenced by postmodernist thinkers such as Foucault and Derrida, argues that the best we can hope for is to work by common rules of civility that will allow people to argue as disputants bound by agreed rules rather than as enemies bent on mutual destruction.

Dryzek takes exception to such lines of thought, arguing that 'there is much more to life and politics than discourses' (Dryzek 2000: 79). Yet in the end his own criticisms of Habermas's ideal speech situation partially echo some criticisms made by the 'difference' democrats. Dryzek, who cites a sizeable body of theoretical work in his support, argues that

Habermas exaggerates the possibilities for consensus. 'Allowance for workable agreements ... in which different participants accept a course of action for different reasons – so long as these reasons have sustained deliberative scrutiny – could quite easily be inserted into the theory of deliberative democracy, and even in the Habermasian edifice in particular' (Dryzek 2000: 48).

Dryzek's wide, discursive, interpretation of deliberation in Western democracy is certainly relevant to the issue of GM crops where all sorts of tactics designed to shock or otherwise mobilise opinion have been deployed. Elections have featured in this process merely as one among many platforms for contending interests, or perhaps we should better say discourses, to advance their projects. However Dryzek is pessimistic about the prospects for developing effective designs that can produce reasoned agreement. This is one of the areas where the Habermasian approach shows theoretical promise. His notion that arenas can be designed in order to produce optimal conditions for fair and effective discourse is something that has sufficient promise to be worthy of pursuit at a practical level. However Dryzek supports Femia's conclusion that while discursive designs such as citizen's juries may be useful in local circumstances 'the larger the scale at which an issue arises, the harder it is to introduce discursive designs to resolve the issue' (Dryzek 2000: 50; Femia 1996: 392–393).

Deliberative polls, which will be discussed in more depth later, involve a representative sample of electors being given the opportunity to discuss the details of a particular issue before declaring their preferences.

Dryzek's work, which gives much theoretical legitimacy to general environmental campaigning, is good news to environmentalists and other campaigners. Tactics such as consumer boycotts are so effective in some cases that, as in the case of GM food policy in many parts of Europe, governmental decisions about GM food and crops have often been rendered irrelevant. The GM food has simply disappeared from retail outlets and, without a market, decisions about whether to license commercial cultivation of crops make little practical difference. Yet this raises a question. How can we make governmental decisions more relevant in such circumstances? It does not seem to increase confidence of an already partially apathetic citizenry in the formal democratic process when the decisions the system produces seem to be irrelevant in the face of the expression of citizen views through non-formal mechanisms.

One option for closing this gap between formal political processes and 'outsider' political activity is to look for practical discursive designs. They could help narrow the gap between (as Dryzek puts it) the contestation of discourses that constitutes practical deliberation in the public sphere and the decision making apparatus of government where decisions arise (as Irwin puts it) as a result of what is usually only a restricted so-called 'science-centred' approach.

Deliberation and discursive designs

There have been popular demands for more deliberative involvement of decisions in the democratic process through, for example, suggestions made by Ross Perot during the 1992 Presidential election (Fishkin 1995: 138–141). These type of suggestions involve increased deliberation by voters as part of the selection process for Presidential candidates, and, presumably, other elected politicians. On the other hand, specific to the environmental field, there has been a general increase in the public availability of environmental information through such mechanisms as the USA's Toxic Release Inventory (Renn *et al.* 1995: 20). As we shall see in Chapter 4 in some ways there is much more involvement of the public built into the US system of environmental regulation than there is in, say, the UK system. What are the options for involvement of lay citizens in deliberation on environmental issues? Perhaps we ought to begin with some deliberative principles articulated by Habermas, who is, after all, the high priest of the importance of reinvigorating the 'lifeworld' in order to counteract trends towards the 'scientisation' of society.

Webler (1995: 46) reports that Habermas's four conditions of the ideal speech situation were:

1   All potential participants of a discourse must have the same chance to employ communicative speech acts.
2   All discourse participants must have the same chance to interpret, claim or assert, recommend, explain and put forward justifications; and problematise, justify or refute any validity claim.
3   The only speakers permitted in the discourse are those who have the same chance to employ representative speech acts.
4   The only speakers permitted in the discourse are those who have the same chance to employ regulative speech acts.

In this context:

> 'representative' speech acts 'claim normative rightness through appeal to legitimate interpersonal relationships (Example: 'Operation of the landfill should be overseen by a committee of elected citizens') Representative speech acts claim sincerity in reference to one's own subjectivity (Example 'I am concerned that the noise from the landfill site will be unbearable').
>
> (Webler 1995: 43)

These explanations are drawn from a rigorous attempt to evaluate different discursive designs for achieving citizen participation put together by Renn *et al.* (1995). They concluded, from experience of designing citizen participation schemes in practice, that such projects could not be evaluated according to the outcomes of planning decisions. They rejected the idea, supported by many administrators, that participation should be judged according to whether it helped secure the agreement for proposals. Conversely, they also rejected the idea that participation should be evaluated according to whether the proposals could be more effectively blocked. Instead the model developed by Renn *et al.* (1995) focused on whether the deliberative procedure satisfies criteria for 'fairness' and 'competence'.

The standard of fairness of participation means that 'everybody takes part on an equal footing . . . not only are people provided equal opportunities to determine the agenda, the rules for discourse, and to speak and raise questions, but also equal access to knowledge and interpretations' (Webler 1995: 38). Fairness implies that all actors are free to attend, initiate discussions, contribute to discussions and make decisions. Competence is assessed according to the degree that procedures about discourse are applied. Four types of discourse discussed by Habermas are considered. These discourses are; explicative, concerned with understanding of meanings and terms; theoretical discourse which is about theories concerning the way nature or the human world operates; practical discourse which is about how social relations are or should be organised, and therapeutic discourse which is about the authenticity and sincerity of claims that are being made (Webler 1995: 66–71).

The practical operations of eight types of citizen participation were evaluated according to the model discussed. These included Citizen Advisory Committees, Citizen Panels, Citizen's Juries, Citizen Initiatives, Negotiated Rule Making, Mediation, Compensation and Benefit Sharing and the Dutch Study groups. Other types of traditional participation such as public hearings, scientific advisory groups or arbitration were ruled out because they met few of the fairness and competence criteria. Referendums were considered to be an effective means of legitimising decisions rather than discussing them (Renn *et al.* 1995: 10).

Some of the types of citizen participation considered, including Citizen's Initiatives, Negotiated Rule Making, Mediation and Compensation and Benefit Sharing, are not relevant to the GM food and crop issue, either because some types of participation are specifically tailored to dealing with local issues or because they are oriented to situations where all interested parties are associated with easily identifiable interest groups. A key feature of the GM food issue is that all of the lay public is potentially affected.

The Dutch Study Group method is one possible application to the GM food/crops issue given that it involves a carefully organised national debate with opportunities for all interested parties to circulate their detailed views which are discussed by a great number of study groups to which there is open access. The results of the discussions are collated and reported. The Dutch Study Group approach was embodied in the national debate (in The Netherlands) about energy policy in the 1980s. In fact the conclusions of the final report were eventually put into practice, but only after the government, having rejected the report findings in favour of a nuclear power station construction programme, were forced to back down by public hostility.

The author of the evaluation concludes that 'such an approach can only be reserved for some very basic strategic social issues and on a low frequency ... The model may be transferred to other fundamental issues of a comparable status and size as the energy problem, for example the issue of genetic engineering' (Midden 1995: 318–319). A key problem is the cost of running such an exercise, estimated at around 250 million dollars in the USA or about 30 million pounds in the UK. In addition there is the obvious problem of political effect. One can design the most deliberatively democratic system imaginable, but it will only be effective if it becomes the norm that such devices are accepted as having legitimacy in the decision making process (Mumpower 1995: 327).

In terms of evaluation according to the criteria set out by Renn *et al.* (1995), the Citizen's Advisory Committees design came out well on competence and fairness criteria for its members, but was handicapped by its very limited access because of the carefully selected membership. Dutch Study Groups came out best on access (since it was completely open), and well on several other criteria. On the other hand, because of the need to have prior organisation of the agenda, the Dutch Study Groups design came out behind the Citizen's Advisory Committees on fairness in deciding agendas and selecting the moderator who organises the terms of discussion. Citizen's Juries involve random selection of citizens who then carefully learn, investigate and cross examine witnesses before making an adjudication. Citizen's Juries do reasonably well on access, although they can lack control over the agenda and may not be able to fully investigate the authenticity of statements made to them in evidence.

One method that was not investigated by the Renn *et al.* (1995) study were Deliberative Polls, which have been popularised by Fishkin (1991, 1995). Deliberative Polls involve the selection of a representative sample of the population to examine, debate and finally vote upon a number of pre-arranged propositions. As Fishkin puts it:

> A Deliberative Opinion Poll gives to a microcosm of the entire nation the opportunity for thoughtful interaction and opinion formation that are normally restricted to small-group democracy ... Most importantly it offers a face-to-face democracy not of elected members of a legislature, but of ordinary citizens who can participate on the same basis of political equality as that offered by the assembly or the town meeting.
>
> (Fishkin 1991)

Fishkin points out that the poll is prescriptive, not predictive. In fact, such polls have been held on a variety of issues in the USA, although they have only very rarely had an effect on policy decisions. However, in theory, an adapted version of the deliberative poll could answer some of the practical problems associated with Citizen Juries and the Dutch Study Groups. Even quite a large Deliberative Poll on GM food and crops could be much cheaper than the Dutch Study Group method, yet the discussions and investigations could approach the thoroughness of the Citizen's Jury Method. On the other hand a much broader number of people could be involved compared to a Citizen's Jury.

What we can say, at the conclusion of this survey of the theoretical possibilities for applying deliberative democratic techniques, is that there are great potentials for applying the theory to the GM food and crop debate. A big problem with the GM crop debate is that there is a shortage of efforts to cross an apparent divide between a scientific/technical elite which is naturally inclined to take a positive view to the impact of scientific and technical advances and a lay-public (especially in the EU) that has its doubts about certain technologies and which cannot be forced to accept the word of the scientists. Hence there would seem to be a desperate need to develop systems of public deliberation that treats these different outlooks on science and technology with equal respect. So let us now move on to investigate the efficacy and extent of efforts to involve the wider public in deliberation over GM food and crop policy in some case studies. First, let us look at the UK's debate about the future of GM food and crops.

Public deliberation and GM food in the UK

The efforts by UK government to set up public deliberation in GM food and crops occurred only after the issue became one of major controversy. Following the setting up of the Farm Scale Evaluation Trials, and about the same time as the reorganisation of ACRE (that is at the beginning of 1999), the government set up the Agriculture and Environment Biotechnology Commission, or AEBC for short. Its objectives are to give 'strategic advice on developments in biotechnology and their implications for agriculture and the environment. It will look at the broad picture taking ethical and social issues into account as well as the science.' (AEBC 2003).

The AEBC consists of 20 members which include representatives of both prominent pro-biotechnology interests and also GM crop sceptics. However the majority of the members are independent experts whose presence guarantees a general agreement with the thrust of government policy which, in its post 1998 formulation, stresses cautious support for GM food and crop technology subject to prescribed environmental tests such as Farm Scale Evaluation Trials. The AEBC has produced some worthy reports on the Crop Trials (AEBC 2001); on future prospects for agricultural biotechnology (AEBC 2002a) and also on Animals and Biotechnology (AEBC 2002b) and the body advised the government how to structure its debate on GM crop regulation. However, it is difficult to pin down any precise changes in government policy that have been produced by the deliberations of the AEBC.

Given that many members of the AEBC act as role-model representatives for conflicting pro- and anti-GM crop interests and causes, there is never going to be a consensus on key points. The discussions conducted at the AEBC are unlikely to come anywhere near the Habermasian ideal of free and unfettered debate where people will change their views according to the strength of superior argument. Moreover, as the government itself stresses, specific power to advise ministers on the issue of authorisations for food and crop tests or commercialisation remains with committees like the Advisory Committee on Releases into the Environment (ACRE). It seems that, apart from its public relations function, the main function of the AEBC is to try to civilise debate. This involves turning protagonists on the issue into what Mouffe (1995) would call agonistic opponents who abide by a common, civilised set of debating rules, rather than antagonists bent on the other's destruction. That is a reasonable aim in itself, and it is probable that the AEBC does contribute towards this aim, as well as providing a good source of well-checked information. The AEBC certainly struggled hard to try to persuade the government to hold a thoroughgoing public debate on the GM food and crops issue. However, it seems that the government did not fully appreciate the efforts of the AEBC in this direction.

Even the AEBC's recommendations on how to structure the debate about GM crops that the government organised in the 2002–2003 period were only patchily implemented by the government. Three key criticisms of the structure of the debate can be located. The first is the scale of the resources, which were far too limited for a meaningful exercise. The second was the total absence of anybody from the public on the AEBC, the Science Review Board or the Steering Committee of the Public GM debate. In a GM food and crop debate that is conspicuous by the apparent division of interpretations of risk between experts and lay public it seems that restricting meaningful discussion to the experts once again is merely perpetuating this key problem of division without meaningful communication. In theoretical terms it also perpetuates the mistaken belief, as discussed in Chapter 2, that the 'science' of GM crops and food can be divorced from the belief systems that influence opinions about the desirability of the technology and its various specific applications. The third key flaw is the refusal, in advance, by the government to accept the need for the public to give a clear message about key, specific, issues in the GM food and crop policy. The AEBC said it wanted to set up 'panels of sixteen or so citizens recruited to discuss the issues in depth' adding a proviso that 'it would be important that the government gave a commitment to take their conclusions seriously in the decision making process' (Grant 2002).

A letter stating government policy said:

> While not wishing to constrain the extent to which the agenda is driven by the public, we would ask the steering board to ensure (wherever possible) that the focus is principally but not exclusively on the issues raised by the possible commercialisation of GM crops. As indicated in your advice, the main objective of stimulating public debate is to establish the nature and spectrum of the public's views on the possible commercialisation of GM crops in this country. As you recognise, Ministers are not seeking a quasi-referendum on the future of GM crops and would have genuine difficulties in taking account of a final report which was quantitative rather than qualitative in its substance.
>
> (Finney 2002)

The government decided to have a three track review: a public debate; a science review; and an economics review. The science review was an expert review, which, given the balance of experts, would produce broadly predictable views that were in keeping with established government policies. The science review panel consisted of 17 professors, seven PhDs and only one other, Julie Hill, from the Green Alliance. There were certainly no ordinary people here even if a couple were classed as 'lay members'. The balance of the experts was clearly favourable towards existing government policy of support for commercialisation of GM crops subject to the herbicide tolerant crops demonstrating, in the Farm Scale Evaluation Trials, that they would not damage biodiversity. The economics review was no different from a classic consultation exercise whereby submissions were invited and a report would be written-up by government.

The public debate promised something different in that it was called, and in some very limited respects was, 'deliberative'. Nine regional workshops were held of grass-roots people to 'frame' the issues for the later national debate. Some focus group research was commissioned. However, even though funding for the debate was extended to £500,000, this was widely felt to be very inadequate for the purposes of organising a national debate which involved the wider British population. The national debate was organised by a Steering Committee involving mainly members of the AEBC in rough proportion to the balance of views represented on the AEBC. A series of public meetings were organised and responses were invited from the public in a debate called 'GM Nation?' In the end the failure to organise a well structured deliberative exercise backfired on the government. When the AEBC issued its report on its public consultations (in September 2003) it found that overwhelming majority of the responses were hostile to GM food. Pro-GM food interests claimed that the debate had been hijacked by well organised anti-GM campaigners. However, the lack of a rigorous deliberative exercise meant that such claims could not be tested.

The nearest thing in the public debate to an exercise in public deliberation (other than consultation meetings) was a 'Citizens' Jury' exercise. A polling organisation, Opinion Leader Research (OLR) was commissioned by the Food Standards Agency (FSA) to put the Citizens' Jury idea into practice. However, the terms of this exercise are open to some serious criticisms in terms of the rules for deliberative design discussed earlier in the chapter.

The Jury consisted of 15 people drawn from a cross-section of the population in Somerset. There was then a series of well-organised presentations and discussions. This part was well done. Other aspects were more controversial. A key procedural problem in the Jury set-up was that the central policy options that were presented to the Jury members were open to differing interpretations, and did not really focus on specific issues of policy contention. The OLR and the FSA, in their reports and press releases, focused on one part of the questionnaire as the central issue. In this part the Jury members were asked to agree with one of four statements:

1   All GM food should be freely available to buy in the UK.
2   All GM food should be freely available to buy in the UK but clearly labelled and tightly regulated.
3   All GM food should not be freely available to buy in the UK until further tests have been completed.
4   GM food should never be available to buy in the UK.

Option 1 received one vote. Option 2 received eight out of 15 Jury votes Option 3 received six votes and Option 4 received no votes (OLR 2003: 87). Opinion Leader Research, and the Food Standards Agency summarised the result in these terms:

> A majority of nine Jurors concluded that GM food should be available to buy in the UK. A sizeable minority of six Jurors said GM food should not be available to buy in the UK.
>
> (OLR 2003: 5; FSA Press Release 2003)

As can be seen, this 'headline' summary omitted the key phrase 'but clearly labelled and tightly regulated' from the statement which the majority actually supported according to the published outcome (OLR 2003: 87). This latter point was seized upon by Friends of the Earth (Friends of the Earth 2003). They announced that the Jury had undermined the FSA's previously stated policy on GM food labelling. The FSA, in line with the UK government's position, had opposed the proposed (now approved) EU's Food and Feed and Traceability and Labelling regulations which advocated the labelling of foods derived from GM food products.

The FSA also opposed the lowering of the thresholds of GM food content (as a requirement for labelling as ' may contain GM food') below 1 per cent, again, in contrast to the new EU regulations.

Just about the only clear quantitative result to emerge from the questions posed of the Jury was that when asked whether they had become 'more' or 'less' concerned about GM food as a result of being on the Jury, nine actually said they had become more concerned, five less concerned and one said their view was unchanged (OLR 2003: 86). Given the already sceptical nature of British public attitudes to GM foods, the idea that in depth discussion actually leaves a Jury (organised by a government organisation) more concerned about GM food cannot be comforting for the government.

The Dutch GM debate

The Dutch government followed the precedent set during the debate about nuclear power (described earlier in the chapter) by launching a debate about GM food and crops in 2001. The Dutch cabinet gave the running of the debate to the Parliamentary Commission on Biotechnology and Food. The debate was well funded – around £2.5 million or $4 million was spent on the debate. In per capita terms this was around 20 times the money spent on the GM debate in the UK.

The debate consisted of three layers. An inner layer consisted of 150 ordinary citizens selected as being representative of the population who were broken down into six groups. They conducted extensive discussions in the context of considerable resources and access to detailed information from a variety of opinions. A second layer consisted of a series of public debates around specific topics and case studies organised at a range of local organisations. A third layer was a more general and wide-ranging public debate including media discussions, videos and public debates held in different localities (Terlouw *et al.* 2002).

At the end of the debate the Commission made three general recommendations: for the setting up of a European Food Authority that is independent of business and government; the organisation of early dialogues with the public on the topic of biotechnology; the provision of full product information to the consumer. Although the debate was well supported, and received very widespread coverage in the media, it's procedure was criticised by anti-GM food groups such as Greenpeace. In particular they complained that the debate was organised around the theme 'under what conditions is the use of modern biotechnology in food (production) acceptable?' rather than 'is the use of modern biotechnology in food production acceptable?' They also complained that key GM policy decisions had already been taken.

As was the case in the earlier Dutch Study Groups public debate on nuclear power, the agenda and questions to be considered in the Dutch

GM debate were decided before the representative population samples (in the first layer of the debate) were selected. This limits the ability of the public to engage in fully open discourse, and it also reduces the legitimacy of the exercise to the extent that it opens up the process to attack from those who object to the way the agenda is set. However, it does seem that the aims of encouraging much wider debate and knowledge about the issue were achieved (Terlouw *et al.* 2002), and on this basis, the Dutch GM debate scores more highly than the elitist and even more carefully-managed British GM debate.

Official debates about GM food have followed the contours of statutory consultative procedures. Let us have a look at the nature of the public debates on aspects of GM food policy that have taken place here.

Public deliberation in the USA

The debate about GM crops and food in the USA has involved rather more consistent public involvement in formal regulatory deliberations than has occurred in the UK. This is not because of any greater public concern about the topic in the USA – concern has generally been a great deal stronger in the UK and the EU as a whole – but rather because of the general requirements laid down in US regulatory practice covering food, health and environmental issues. Under US regulatory procedure, every time there is a proposal for a new rule, or even a petition for a new GM product, then public comment is invited as part of the discussion which precedes a risk management judgement by the FDA, EPA or USDA. Judging by the number of people submitting comments there has been a sharp increase in the number of people interested in the GM crops/food issue. As we have seen from the earlier discussion, comments submitted to the Calgene (Flavr Savr Tomato) case in 1992–1993 numbered dozens, but in 1999 comments to the proposed FDA rules on GM food safety and labelling numbered several tens of thousands.

That having been said, the debates that do occur are far from the sort of open-ended debates about values that critics of GM food and crops want. Open-ended debates about values are the sort of discussions that deliberative theorists such as Renn *et al.* (1995) might suggest as being necessary. However, the dominant discourse among the scientific elite seems to be that while public consultation is important, the detailed decisions are best left to the experts:

> It is clear that democracy is best served when people affected by regu-latory decision making can be significantly involved in the decision making, and that inclusion of diverse interests in the risk analysis process can be a powerful force to garner legitimacy of a decision. This is especially true because the significance of environmental effects of novel genetic material depends on societal values, However,

especially when the decision options under consideration are not well defined, broad public involvement in risk analysis can result in risk management decisions that lack scientific rigour.

(National Academy of Sciences 2002b: 6)

In this sense the National Academy of Science seems to adopt at least some of what was earlier described as a 'science-centred' view which precludes the involvement of the public in decisions in what are said to be technical matters (Irwin 1995: 14). It is an approach which the establishment can deploy with less resistance in the USA compared to EU countries. Nevertheless, it would be churlish to ignore the longstanding statutory consultation procedure. This can allow effective mobilisation of opinions to counter entrenched industrial interests, as can be seen in the case of setting national organic farming and food standards.

Lynch and Da Ros (2001) analysed the FDA public meetings held to discuss possible changes in food safety and labelling rules. The format of the public meetings consisted of discussions by two panels (on GM food safety and labelling) and an opportunity for some members of the public to make statements. As Lynch explains:

'The first panel focused on scientific and safety issues, and consisted of university and industry scientists. The second panel (on labelling and public information) was composed of representatives from academia, non-governmental organisations, and industry groups representing both organic and conventional farming.' Only the FDA was able to question panelists directly.

(Lynch and Da Ros 2001)

As was clear in earlier years, the scientific advisers on the panel wanted to increase the requirements on biotechnology companies to demonstrate the safety of GM food. The panel on labelling backed a policy of voluntary labelling. This was broadly in keeping with the changes that the FDA later made, that is the FDA's adoption of the notion of mandatory notification of the safety (although not actual scientific assessment) of GM food and of voluntary labelling.

Given that the FDA had the power to appoint a balance of representatives who would not violently disagree with the general thrust of FDA policy it is not surprising that policy ended up supporting the FDA's general views. As in the case of the UK, both scientific and 'participatory' advisory mechanisms are constituted to reflect the concerns of the dominant policy network which, in any case, forms the value basis upon which policy decisions are taken and scientific assessment procedures care chosen.

Conclusion

In the UK case I have examined the UK public debate on GM food and crops and the role of the Agriculture and Environment Biotechnology Commission (AEBC) as a means of developing deliberative democracy. The AEBC itself has generated some degree of deliberation, but its main function has been to improve the 'civility' of the debate. Much of its efforts to achieve a genuine deliberative debate which produces a public discussion and verdict on key regulatory issues have been thwarted by the government. Although the Food Standards Agency did appear to conduct a Citizen's Jury Debate in answer to a call made by the AEBC, the actual Citizens' Jury falls a long way short of the principles of deliberative design discussed earlier in this chapter. The Dutch GM debate scores well on the objectives of achieving widespread public involvement, however, there were still important divisions on the framing of the way that the debate and the key questions that were to be addressed.

By contrast, the public deliberation on the GM food and crops issue in the USA has been channelled through existing statutory provisions for public consultation. These have given possibilities for interest groups such as the organic food lobby to defend their interest (as described in Chapter 4), and there have been hearings on issues such as GM food labelling. However there has been no attempt to organise a deliberative, as opposed to consultative, discussion on GM food and crop issues. The public is only represented 'virtually' by experts sitting on scientific advisory panels dealing with questions framed by government Agencies.

In all of the cases there is a great reluctance on the part of governments to establish mechanisms such as deliberative polls or citizens' juries which come to conclusions about precise regulatory issues. Governments argue that they have been elected to govern and do not appear to want to legitimise an alternative centre of power. Yet, in the long run, they may find that they are impelled by force of public opinion to adopt the type of decisions that would be thrown up by deliberative mechanisms. Governments still have the responsibility to govern, but they may find there are advantages in allowing the articulation rather than the suppression of popular views. This could, ultimately, save governments a lot of political problems over political issues such as GM food and crops. For deliberative systems to work the representatives who are selected to reflect the public view need to be given more room to vary the way that questions are framed given that such deliberative mechanisms that do take place are often given either over-generalised or ambiguous questions to answer. Of course, even then there is anther problem in that public attitudes change, so one deliberative verdict cannot be expected to end debate for all time.

Conclusion

The most important task of this book has been to analyse how political outcomes on GM food and crops have diverged in the USA compared to the EU and the UK. I have warned, in Chapter 2, against a search for definitive causes of outcomes. It would be wrong simply to ascribe outcomes to specific events or specific actions by agents. What we can do, however, is discuss contexts and events that can be associated with differing outcomes, while at the same time taking care to recognise that actions by agents and events will be interpreted differently in different contexts.

Notwithstanding these caveats, I find that three factors are especially associated with the differences between outcomes in the USA and the EU. First, the different attitudes to food that are prevalent in the USA compared to Europe. Second, that the environmentalist discourse concerning corporate control of allegedly dangerous technologies such as GM food and crops has much greater resonance with European rather than US publics. A third factor has been the BSE crisis which may have reinforced European, especially British, suspicion of GM food. I shall expand on these three factors.

A key finding is that the differences in outcome are associated with different attitudes to food in the USA compared to Europe and also with a greater scepticism towards the opinions of official scientists in Europe compared to the USA. In the USA food is more often regarded as a fuel which is consumed quickly and valued for its relative cheapness. Europeans are more likely to value food for aesthetic qualities such as naturalness and will often put such qualities ahead of cost as a criterion for purchasing choice. Europeans, especially in countries such as Italy, France and Greece, tend to see food as a part of their own national identities. Appeals to the notion of their identity being undermined by what are seen as 'Americanised' food products, which are bereft of association with either history or specific place of origin, have had widespread resonance.

Environmentalist opposition to what they see as corporate control of potentially dangerous GM food and crop technology has had widespread sympathy within countries such as Germany, Austria and Denmark since the 1980s. This suspicion of GM technology may well be associated with the criticism of the nuclear power industry in these countries given the similarity of discourses about undemocratic corporate control and the alleged abuse of science and technology for the benefit of large corporations. In the USA, by contrast, there has, since the Reagan era of the 1980s, been much greater tolerance for the activities of 'corporate' power. There has, in the USA compared to the EU, also been a much greater faith in the ability of regulatory institutions to work in a free market context to ensure adequate environmental and health protection.

The BSE crisis has been widely cited as a factor in explaining US–EU differences over the issue. As I have argued in Chapter 5, the BSE crisis seems likely to have served to reinforce existing suspicions of GM food and crops rather than create them. There is evidence that the BSE crisis was influential in the UK in reinforcing anti-GM food attitudes, although even here, as elsewhere in Europe, it is difficult, if not impossible, to separate the effect of the crisis itself from the use of the crisis as a new metaphor for food safety fears. I certainly think that the impact of the BSE crisis on attitudes to GM food can be exaggerated. Indeed it is plausible to argue that even if the BSE crisis had not occurred at all, the EU and the USA would still have ended up in dispute over the issue, although the process might well have taken longer to develop. Certainly, if the BSE crisis had occurred in the USA it is not at all clear that this would have led to US consumers rejecting GM food in the same way as the Europeans.

These factors have been associated with different dominant regulatory discourses and rules on issues such as labelling of GM food. The fundamental difference in regulatory discourse is over whether GM plants are different to non-GM plants. The EU system asserts that genetically engineered plants are not 'natural' while dominant US regulatory discourses assert that genetic engineering involves nothing that is fundamentally different from traditional methods of improving characteristics through plant breeding techniques. These conflicting discursive positions are associated with different conclusions about whether GM food should be labelled or not. The labelling of GM food has been the crucial action that has effectively isolated GM food products from EU and other food markets. Increasingly farmers around the world have seen it as being in their interests to produce non-GM food that will be accepted by Europeans and others who refuse to buy GM food. Far from, as feared earlier, Europeans having to accept GM food because of the lack of alternatives, farmers in places like Canada are now opposing the granting of authorisations to new GM crops such as herbicide tolerant wheat for fear that they will lose export markets.

The differing types of scientific assessment of GM food and crops in the USA compared to that of the UK and the EU are related to different cultural attitudes. The UK scientific establishment supports a more precautionary approach to GM food and crops, than say, the US scientific establishment. This inter-relation of scientific opinion, cultural attitude and regulatory practice constitutes distinctive discourse-paradigms on GMOs in each country. It is worth repeating the conceptualisation (set out in Chapter 4) of the discourse-paradigm approach as represented by the example of GM food and crop policy:

1   How judgements about what are seen to be scientific issues relating to GM food and crops are done in relation to a distinctive set of regulatory discourses which are linked to culturally based attitudes.
2   How scientists, who are informed by these commonly held cultural attitudes, are imbued with a distinctive set of attitudes about issues of controversy on agricultural biotechnology.
3   How these sets of attitudes produce a distinctive set of 'facts' which needs to be collected.
4   How the scientists involved in scientific advisory work are broadly representative of attitudes to GM food and crops held by their respective national scientific 'peak' organisations, in these cases (the USA and the UK) the National Academy of Sciences (NAS) in the USA and the Royal Society in the case of the UK.

In the case of the EU the European Parliament (although not the Commission) is influenced by a series of nationally based discourse-paradigms to produce a composite political verdict. It has been the European Parliament, in collaboration with dissident member states, that has acted as the transmission belt for popular scepticism to GM food and crops. The 'two-level' system of regulation of GM food and crops has helped sceptics of GM food and crops to make their moratorium of GM authorisations more effective. However it should be emphasised that it is the pressures of the consumers to boycott GM produce sold in shops and also the popular pressures through the European Parliament on the legislative front that seem to be the crucial modes of influence in this case study.

It is certainly true that the BSE crisis has been associated with greater hostility towards GM food in the EU. However, one should be wary of interpreting events as though their effect is the same regardless of cultural context. The European and the (USA) Americans are likely to construct, to interpret, the same events rather differently. This is especially the case given the greater sensitivity of Europeans (compared to US citizens) to food quality issues and food 'crises'. In short, we need an 'interpretivist' rather than positivist approach to understanding these comparisons.

The USA and the EU use different set of scientific standards to evaluate GM food and crops. So who is being unscientific? Is it the EU or the USA? Perhaps this is a bad question, for much of what is called science in this GM food and crops debate is in fact, culture. Given the differing priorities held by the USA and the EU, and the existence of uncertainties, a search for absolute 'truth' may be unfulfilled. The controversial assessments of the risks of using antibiotic resistant marker genes, the issue of whether

GMOs are fundamentally different to conventional crops, of whether we should subject GM food to scientific assessment as opposed to just demonstrate their substantial equivalence to conventional food, the issue of what tests we should select to determine the impact on wildlife by GM crops; these are all 'value' issues. They are not matters of natural science – or indeed any other sort of science – that can be settled (as opposed to informed) by recourse to positivist exercises in hypothesis testing. Much, if not most, of what passes for 'science' in the debate about GM crops and food is in fact nothing more than a more technical elaboration of dominant cultural values which act as social truths of the day in a particular country.

I must confess that when I began my research into GM food and crops with a study of UK controversies on this subject my focus on the British concern with impact of GM crops on the food supply of birds persuaded me that differences in agricultural practice in the USA and the EU were the main cause of the differences in regulatory approach. In the USA, as opposed to Europe, only a minority of the land surface is farmed in some form, thus reducing concern about impact on birds. I can now see that the prominence given to the 'food supply for birds issue' in the UK is a feature of institutional path dependence, not a key explanation for the policy differences. Bodies like English Nature and the RSPB dominate governmental environmental policy networks, so when the GM food and crop issues became matters of high consumer and public concern it was interpreted according to the pre-existing frame of dominant environmental groups. However, the key issue, in the UK and the EU, is the level of consumer food concern, as well as the general suspicion of GM technology by environmental groups at large and the impact of the BSE crisis.

The central problem with taking the dispute over GM food to the WTO is that the WTO procedure is designed to deal with cases of alleged trade protection. Yet, as I have argued in Chapter 5, trade protection has little to do with the EU's refusal to consume GM food products and license GM crops. The European farmers have been, by and large, quite keen to take advantage of the technology. It is consumers, organised by environmental groups with their own agenda, that have prevented the European Commission from avoiding this dispute. The complaint to the WTO is unlikely to help US farmers. Whatever the outcome of the WTO dispute, the USA is losing its attempt to persuade the world to accept GM food, and in doing so its own farmers are suffering.

Interviews

Please note that dates are given in English rather than US style i.e. day/month/year.

Abramson, Stanley (former EPA official, now partner to Arent Fox Kintner Plotkin and Kahn, PLLC) 24/02/03.

Barber, Simon, EuropaBio, 19/07/02.

Baynton, Clare, Food Standards Agency (UK), 02/08/01.

Case, Jeffrey, Corporate State Affairs Lead, Syngenta, 25/02/03.

Consoli, Lorenzo, European Political Advisor on Genetic Engineering for Greenpeace, 19/07/02.

Curtoys, John, Public Affairs Co-ordinator, Royal Society for the Protection of Birds (RSPB), 20/04/01.

Devereux, Clare, Co-ordinator, Five Year Freeze Campaign, 01/05/01.

DG Environment Official –European Commission, 17/07/02.

DG Environment Official – European Commission, 29/07/02.

DG Environment Officials –European Commission, 10/06/03.

DG SANCO Official – European Commission, 17/07/02.

Dobson, Conor, European Public Affairs Manager for Bayer (formerly Aventis), 17/07/02.

Environmental Protection Agency Officials, Office of Pesticide Programs, 14/02/03.

Gall, Eric, Greenpeace European GM campaigner, 13/06/03.

Gibbons, David, Chief Scientist, Royal Society for the Protection of Birds (RSPB), 07/08/02.

Giddings, L. Val, Vice President for Food and Agriculture, Biotechnology Industry Organisation (USA), 18/02/03.

Hill, Julie, Former Director of Green Alliance and member of Advisory Committee on Releases into the Environment ACRE (UK), 23/05/01.

Holden, Patrick, Director of Soil Association (UK), 06/06/01.

Jackson, Caroline, MEP, Chair of the Environment Committee of the European Parliament, 04/06/03.

Johnson, Brian, Head of Biotechnology Advisory Unit, English Nature, 16/05/01.

Kettlitz, Beate, Bureau Europeen des Unions de Consommateurs (BEUC), 12/06/03.

Kochenderfer, K., Public Affairs Director, Grocery Manufacturers of America, 20/02/03.

Kronick, Charlie, Genetic Engineering Campaigner, Greenpeace (UK), 08/06/01.

Leskien, Dan, Adviser on Genetic Engineering to the Green European Free Alliance, 13/09/02.

Lewandowski, Magda and Ribera, Marie-Christine, COPA-COGEMA, 11/06/03.

Margulis, Charles, US Greenpeace Campaigner, 02/03/03.

Mayer, Sue, Director, Genewatch, 18/06/01.

Mendelson, Joe, Senior Advocate, US Center for Food Safety, 27/02/03.

Merritt, Colin, Biotechnology Development Manager, Monsanto UK, 04/07/01.

Mitchell, Larry, Chief Executive Officer, American Corn Growers' Association, 19/02/03.

Morris, John, Food and Drinks Executive of the British Retail Consortium, 29/08/01.

Moses, Vivian, Visiting Professor of Biotechnology, Kings College London, and Chairman of CropGen, 14/05/01.

Muldoon, Joe, Public Affairs Manager, Syngenta, 27/02/03.

Pearsall, Dan, Secretary to Supply Chain Initiative for Modified Agricultural Crops (SCIMAC), 26/06/01.

Piersma, Auke Mahar, Aide to Congressman Dennis Kucinich, 28/02/03.

Plan, Damien, Public Affairs Executive (Brussels) for Bayer Crop Science, 12/06/03.

Riley, Pete, 'Real Food' Campaigner for Friends of the Earth UK, 24/05/01.

Rissler, Jane, Senior Staff Scientist, Union of Concerned Scientists, 26/02/03.

Rootes, Chris, Director, Centre for the Study of Social and Political Movements, University of Kent, 10/04/03.

Scheider, Russell, Director, Regulatory Affairs, Monsanto, 28/02/03.

SCP Scientist, Scientist member of EU Plant Science Advisory Committee, 17/09/02.

Senior Civil Servant at the Department of Environment, Transport and the Regions (DETR) dealing with GM crops, 01/06/01.

Senior Civil Servant at the Ministry of Agriculture, Food and Fisheries (MAFF), 12/06/01.

Senior Former Civil Servant, Ministry of Agriculture, Food and Fisheries (MAFF), 05/06/01.

Smith, Stephen, Managing Director of Syngenta Seeds, 26/06/01.

Taeymans, Dominic, Director, Science and Regulatory Affairs and Dieu, Thierry, Communications Manager, Confederation des Industries Agro-Alimentaires de l'UE (CIAA), 11/06/03.

Watkins, Rosemary, Senior Director, Congressional Relations, American Farm Bureau, 02/03/03.

Whitehead, Geoffrey, MEP, 11/06/03.

Williams, Ingrid, Professor of Entomology at the Institute of Arable Crop Research and member of the Advisory Committee on Releases into the Environment (ACRE), 28/08/02.

Bibliography

Abramson, S. and Carrato, J. (2001) 'Crop Biotechnology: The Case for Product Stewardship', *Virginia Environmental Law Journal*, volume 20, no. 2: 241–266.

Advisory Committee on Novel Foods and Processes (2001) *Annual Report 2000*, London: Food Standards Agency.

Agriculture and Environment Biotechnology Commission (2001), *Crops on Trial*, London: Department of Environment, Food and Rural Affairs.

Agriculture and Environment Biotechnology Commission (2002a), *Looking Ahead – An AEBC Horizon Scan*, London: Department of Environment, Food and Rural Affairs.

Agriculture and Environment Biotechnology Commission (2002b), *Animals and Biotechnology*, London: Department of Environment, Food and Rural Affairs.

Agriculture and Environment Biotechnology Commission (2003), 'What the AEBC Does', London: Department of Environment, Food and Rural Affairs, http://www.aebc.gov.uk/aebc/about_us.html, accessed 28/01/03.

Anderson, L. (1999) *Genetic Engineering, Food, and Our Environment – A Brief Guide*, Totnes, Devon: Green Books.

Bailey, R. (2002) *GM Trade War*, National Post 12th August 2002, http://www.truthabouttrade.com/1071/wrapper.jsp?PID=1071–22&CID=1071–081202A, accessed 10/12/02.

Baumann, Z. (1992) 'The Solution as a Problem', Times Higher Educational Supplement, 13th November, p. 25.

BBC (2002) 'Freak Wave', Transcript of *Horizon* programme broadcast on BBC on 14th November, http://www.bbc.co.uk/science/horizon/2002/freakwavetrans.shtml, accessed on 05/12/02.

Beck, U. (1998) *Risk Society – Towards a New Modernity*, London: Sage.

Beringer, J. (1998) *Annual Report of Advisory Committee On Releases into the Environment (ACRE) No. 4: 1996/1997*, London: Department of Environment, Transport and the Regions, http://www.defra.gov.uk/environment/acre/annrep4/1.htm, accessed 31/12/02.

Beringer, J. (1999) *Annual Report of Advisory Committee On Releases into the Environment (ACRE) No. 5: 1998*, London: Department of Environment, Transport and the Regions.

Bevir, M. (1999) 'Foucault, Power and Institutions', *Political Studies*, volume XLVII, pp. 345–359.

Bevir, M. and Rhodes, R. (2002) 'Interpretive Theory' in Marsh, D. and Stoker, G. (eds) *Political Theory and Methods*, London: Palgrave Macmillan, pp. 131–152.

Bourdieu, P. (1993) *Sociology in Question*, London: Sage.

Brown, P. (1999) 'Minister Orders GM Watchdog Clearout', *Guardian*, April 12th, p. 9.

Brown, P. (2003) 'GM Foods: Unloved, Unwanted and A Rush to Grow Crops Could Cause Civil Unrest', *Guardian*, July 12th, p. 4.

Business Science and Technology (1979) 'Genetic Engineering – Improving Plants', *The Economist*, November 3rd, pp. 72–73.

Caplan, P. (1997) 'Approaches to Food, Health and Identity' in Caplan, P. (ed.) *Food, Health and Identity*, London: Routledge, pp. 1–31.

Carson, R. (1962) *Silent Spring*, New York: Fawcett Crest.

CBS News (2002) 'Cancer Alert: Bread and French Fries', CBS News, April 25th, http://www.cbsnews.com/stories/2002/04/25/health/main507222.shtml, accessed 23/05/03.

Charles, D. (2001) *Lords of the Harvest*, Cambridge, Massachusetts: Perseus Publishing.

Checkbiotech (2001) 'Modified Crop Advocates Fear Lost Chances Through Cultivation Ban', *South China Morning Post*, April 18th, www.checkbiotech.org/root/index.cfm, accessed 17/12/02.

Clover, C. (2002) 'Worst Ever GM Crop Invasion', Telegraph Newspapers online, filed 19/04/02, http://www.portal.telegraph.co.uk/news/main.jhtml?xml=/news/2002/04/19/wgm19.xml&sSheet=/news2002/04/19/ixnewstop.html.

Collins, H. and Pinch, T. (1994) 'Representativeness and Expertise: A Response', Public Understanding of Science, volume 3, pp. 331–337.

Collins, H. and Pinch, T. (2000) *The Golem: What You Should Know About Science*, Cambridge: Cambridge University Press.

Collins, H. and Yearley, S. (1992) 'Epistemological Chicken' in Pickering, A. (ed.) *Science As Practice and Culture*, Chicago: University of Chicago Press, pp. 301–326.

Comite des Organisations Professionelles Agricoles de l'Union Europeenne/ Comite General de la Cooperation Agricole de l'Union Europeenne (2002), *COPA and COGECA Remarks on the Draft Regulations of the Council and European Parliament on Genetically Modified Food and Feed and the Traceability and Labelling of Genetically Modified Organisms*, Brussels: COPA/COGECA.

Commission of the European Communities (1986), Revision 1 (23/05/86) draft of: *A Community Framework for the Regulation of Biotechnology*, Communication from the Commission to the Council, COM (86), 573, final, 04/11/86 cited by Shackley *et al.* (1992).

Commission of the European Communities (1996), *Report on the Review of Directive 90/220/EEC in the Context of the Commission's Communication on Biotechnology and the White Paper*, COM (96), 630, final, Brussels, 10/12/96.

Commission of the European Communities (2001), *Proposal for a Regulation of the European Parliament and of the Council on Genetically Modified Food and Feed*, COM (2001), 425, final, Brussels, 25/07/01.

Commission of the European Communities (2002), *Life Sciences and Biotechnology – A Strategy for Europe*, COM (2002), 27, final, Brussels, 23/01/02.

Consumers Association (2002a) *GM Dilemmas – Policy Report*, London: Consumers Association.

Consumers Association Press Release (2002b) *Food Standards Agency (FSA) Stands Out Against Consumers*, November 14th, London: Consumers Association, http://www.which.net/media/pr/nov02/general/fsa.html, accessed 09/07/03.

Cram, L. (1997) *Policy-making in the EU*, London: Routledge.

CropGen Press Release (2001) 'Results of NOP Survey into GM Foods', London: CropGen, March 8th.

CropGen (2002) 'Unsigned – Questions and Answers – Don't Dr Pusztai's Experiments with GM Potatoes Show Harm to Rats who Ate Them?, http://193.118.100.80/databases/cropgen.nsf/Questions+and+Answers+Web+View?OpenForm, accessed 24/10/02.

Cummins, R. (2002) *Biotech Bullies: The Debate Intensifies*, http://www.inmotion magazine.com/ra01/geff11.html, accessed 22/10/02.

Derrida, J. (1976) *Of Grammatology*, Baltimore, Maryland: Johns Hopkins University Press, cited by Howarth, D. (2000) *Discourse*, Buckingham: Open University Press, p. 37.

Derbyshire, D. (1999) 'Frankenstein Food Fiasco', *Daily Mail*, February 13th, p. 1.

Dienel, P. and Renn, O. (1995) 'Planning Cells: A Gate to "Fractal" Mediation' in Renn, O., Webler, T. and Wiedemann, P. (eds) *Fairness and Competence in Citizen Participation*, Boston: Kluwer, pp. 117–140.

DiMaggio, P. and Powell, W. (1991) 'Introduction' in DiMaggio, P. and Powell, W. (eds) *The New Institutionalism in Organisational Analysis*, Chicago: University of Chicago Press, pp. 1–40.

Dobson, A. (1995) *Green Political Thought*, London: Routledge.

Dowding, K. (1995) 'Model or Metaphor? A Critical Review of the Policy Network Approach', *Political Studies*, volume XLIII, pp. 136–158.

Dryzek, J. (2000) *Deliberative Democracy*, Oxford: Oxford University Press.

Dunlop, C. (2000) 'Epistemic Communities: A Reply to Toke', *Politics*, volume 20, no. 3, pp. 137–144.

Dunlop, R. (1991) 'Public Opinion in the 1980s: Clear Concerns, Ambiguous Commitment', *Environment*, volume 33, no. 8: 10–15 and 32–37.

Dyer, G. and Griffith, V. (2003) 'Luck, Science, Timing and the Trials of Biotechnology', *Financial Times*, June 10th, p. 16.

Eckersley, R. (1992) *Environmentalism and Political Theory – Towards An Ecocentric Approach*, London: UCL Press.

Edwards, R. (1998) 'The Price of Arrogance', *New Scientist*, 160 (2156): 18–19.

Elkington, J. and Burke, T. (1987) *The Green Capitalists*, London: Gollancz.

Elliot, J. and Iredale, W. (2003) 'Townie Farmers With Taste Produce Britain's Best Food', *Sunday Times*, June 22nd, p. 11.

Emberlin, J., Adams-Groom, B. and Tidmarsh, J. (1999) *A Report on the Dispersal of Maize Pollen*, Worcester: National Pollen Research Unit, University College, Worcester.

EuroCommerce Press Release (2002) *Member States Opt for Enhanced Transparency and Consumer Choice*, Brussels: EuroCommerce, December 2nd.

Euro Coop (2003) 'Euro Coop calls for True Consumer Choice about GM Food', Brussels: Euro Coop, http://www.eurocoop.org/publications/en/position/co-existence03.asp, accessed 13/06/03.

Falkner, R. (2000) 'International Trade Conflicts of Agricultural Biotechnology' in Russell, A. and Vogler, J. (eds) *The International Politics of Biotechnology*, Manchester: Manchester University Press, pp. 142–156.

FAO/WHO (2003) *Draft Principles for the Risk Analysis of Foods Derived from Modern Biotechnology*, Secretariat of the Joint FAO/WHO Food Standards Programme Food and Agriculture Organization of the United Nations, Rome, Italy, http://www.codexalimentarius.net/biotech/en/ra_fbt.htm accessed 10/07/03.

Femia, J. (2000) 'Complexity and Deliberative Democracy', *Inquiry*, 39, pp. 359–397, cited by Dryzek, J. (2000) *Deliberative Democracy*, Oxford: Oxford University Press.

Fernandez-Cornejo, J. and McBride, W. (2000) *Genetically Engineered Crops for Pest Management in U.S. Agriculture*, Agricultural Economics Report No. 786, Washington, D.C.: Economic Research Service of the US Department of Agriculture.

FIFRA Scientific Advisory Panel (2000a) Executive Summary of Biopesticides Registration Action Document: Bt Plant Pesticides (Background Documents) *Issues Pertaining to the Bt Plant Pesticides Risk and Benefit Assessments*, October 1st–20th, Arlington, Virginia: Environmental Protection Agency, http://www.epa.gov/oscpmont/sap/2000/october/brad1_execsum_overview.pdf, accessed 13/05/03.

FIFRA Scientific Advisory Panel (2000b) Science Assessment, Biopesticides Registration Action Document: Bt Plant Pesticides (Background Documents) *Issues Pertaining to the Bt Plant Pesticides Risk and Benefit Assessments*, October 18th–20th, Arlington, Virginia: Environmental Protection Agency, http://www.epa.gov/oscpmont/sap/2000/october/brad2_scienceassessment.pdf, accessed 13/05/03.

FIFRA Scientific Advisory Panel (2000c) Panel Members, *Issues Pertaining to the Bt Plant Pesticides Risk and Benefit Assessments*, October 18th–20th, Arlington, Virginia: Environmental Protection Agency, http://www.epa.gov/oscpmont/sap/2000/october/panel.pdf, accessed 27/05/03.

FIFRA Scientific Advisory Panel (2002), *Transcript of Open Meeting on Corn Rootworm Pant-Incorporated Protectant Non-target Insect and Insect Resistance Management Issues*, August 27th, Arlington: Virginia. Environmental Protection Agency, http://www.epa.gov/oscpmont/sap/2002/august/sap827.pdf, accessed 10/04/03.

Finney, R. (2002) Letter to Malcolm Grant, Chair of Agriculture and Environment Biotechnology Commission, August 19th, http://www.aebc.gov.uk/aebc/public_attitudes_debate_gov_further.html, accessed 28/01/03.

Firchow, P. (1972) *Aldous Huxley – Satirist and Novelist*, Minneapolis: University of Minnesota Press.

Fischer, F. (2003) *Reframing Public Policy*, Oxford: Oxford University Press.

Fishkin, James S. (1991) *Democracy and Deliberation: New Directions for Democratic Reform*, New Haven, CT: Yale University Press.

Fishkin, J. (1995) *The Voice of the People*, New Haven: Yale University Press.

Food and Drug Administration (1992) *Statement of Policy: Foods Derived From New Plant Varieties*, 57 FR 22984 [Docket No. 92N-0139], Washington, D.C.: US Department of Health and Human Sciences.

Food and Drug Administration (1994) *Secondary Direct Food Additives Permitted in Food for Human Consumption*, 21 CFR Parts 173 and 573 [Docket No. 93F-0232], Part II 59 FR 26700, Washington, D.C.: US Department of Health and Human Sciences.

Food and Drug Administration Center for Food Safety and Applied Nutrition (1998) *Guidance for Industry: Use of Antibiotic Resistant Marker Genes in Transgenic Plants – Draft*, Washington, D.C.: US Department of Health and Human Services, http://www.cfsan.fda.gov/~dms/opa-armg.html, accessed 05/03/03.

Food and Drug Administration (2001a) *Premarket Notice Concerning Bioengineered Foods* (proposed Rule) 21 CFR Parts 192 and 592, Washington, D.C.: US Department of Health and Human Services.

Food and Drug Administration Center for Food Safety and Applied Nutrition (2001b) *Guidance for Industry, Voluntary Labeling Indicating Whether Foods Have or Have Not Been Developed Using Bioengineering*, Washington, D.C.: US Department of Health and Human Services.

Food and Drug Administration/ US Department of Agriculture (2002) *Guidance for Industry: Drugs Biologics and Medical Devices Derived from Bioengineered Plants for use in Humans and Animals – Draft Guidance*, Washington, D.C.: US Department of Health and Human Services, http://www.fda.gov/OHRMS/DOCKETS/98fr/02d-0324-gdl0001.pdf, accessed 13/05/03.

Food Standards Agency Media Centre (2000) *Agency's View on Further Information on GM Soyabeans*, London: Food Standards Agency, issued 02/06/00.

FSA Press Release (2003) 'Citizens' Jury Verdict Announced', April 7th, London: Food Standards Agency, http://www.food.gov.uk/news/newsarchive/verdict_due, accessed 25/07/03.

Foucault, M. (1977) *Discipline and Punish*, London: Allen Lane.

Foucault, M. (1980) *Power/Knowledge – Selected Interviews and Other Writings 1972–1977*, Brighton, Sussex: Harvester Press.

Foucault, M. (1994) *The Archaeology of Knowledge*, London: Routledge.

Foucault, M. (1998) *The Will to Knowledge – The History of Sexuality, Volume 1*, Harmondsworth, Middlesex: Penguin.

Frewer, L. (1999) 'Public Perceptions of Genetically Modified Foods in Europe', *Journal of Commercial Biotechnology*, volume 6, no. 2: pp. 108–115.

Frewer, L., Howard, C., Hedderley, D. and Shepherd, R. (1996) 'What Determines Trust in Information About Food-Related Risks? Underlying Psychological Constructs', *Risk Analysis*, volume 16, no. 4: 473–485.

Friends of the Earth Press Release (2003) *GM Jury Challenges FSA Policy on Labelling*, April 8th, London: Friends of the Earth, http://www.foe.co.uk/resource/press_releases/gm_jury_challenges_fsa_pol.html, accessed 25/07/03.

Gaskell *et al.* (2003) 'Europeans and Biotechnology in 2002', *Eurobarometer 58.0 – A Report to the EC Directorate General for Research from the Project 'Life Sciences in European Society'* QLG7-CT-1999–00286, London: London School of Economics, http://europa.eu.int/comm/public_opinion/archives/eb/ebs_177_en.pdf, accessed 15/06/03.

Gaskell, G. and Bauer, M. (eds) (2001) *Biotechnology 1996–2000*, London: Science Museum.

Gill, B. (1996) 'Germany: Splicing Genes, Society', *Science and Public Policy*, volume 23, no 3: 175–179.

Gillis, J. (2002) 'Corn for Growing Far Afield?; A Mishap With Gene-Altered Grain Spotlights the Odds of Contamination', *Washington Post*, November 16th, Financial p. E01.

Goff, E. (1991) 'Gene Genie', *New Farmer and Grower*, issue 31, pp. 34–35.

Goffman, E. (1975) *Frame Analysis: An Essay on the Organisation of Experience*, London: Penguin.

Goldenberg, S. (2003) 'Bush Attack on Europe's GM Barrier', *Guardian*, June 24th, p. 14.

Grant, M. (2002) Letter to Margaret Beckett, April 26th, http://www.aebc.gov.uk/aebc/public_attitudes_advice.html, accessed 28/01/03.

Gray, A. (2001) *Annual Report of Advisory Committee On Releases into the Environment (ACRE) No 7: 2000*, London: Department of Environment, Food and Rural Affairs, http://www.defra.gov.uk/environment/acre/annrep7/pdf/acre_annrpt7.pdf, accessed 03/01/03.

Green, D. and Shapiro, I. (1994) *Pathologies of Rational Choice Theory*, New Haven and London: Yale University Press.

Greenpeace Press Release (1996) 'Greenpeace Names Ships Bringing Illegal Corn to Europe', Brussels: Greenpeace, http://www.greenpeace.org/pressreleases/, accessed 25/05/01.

Greenpeace Press Release (1998) 'Novartis Maize Remains Banned in Austria, Luxembourg', Brussels: Greenpeace, http://www.greenpeace.org/press releases/, accessed 25/05/01.

Greenpeace Press Release (2000) 'Greenpeace Welcomes European Court Ruling on Controversial GE Maize', Luxembourg: Greenpeace, http://www.green-peace.org/pressreleases/, accessed 07/07/03.

Greenpeace Press Release (2003) 'European Parliament Paves the Way for GMO-Free Europe', Strasbourg, France: Greenpeace, http://www.greenpeace.org/ pressreleases, accessed 08/07/03.

Grocery Manufacturers of America Comments (2003) *Comments Submitted Re: Food Industry Comments on Proposed FDA Regulations for Plant-Made Pharmaceuticals*, February 6th, Washington, D.C.: Grocery Manufacturers of America, http://www.gmabrands.com/publicpolicy/docs/Comment.cfm?DocID=1068&, accessed 06/03/03.

Grocery Manufacturers of America Correspondence (2001) *Letter to Honorable Donald Evans, Secretary at US Department of Commerce*, November 6th, Washington, D.C.: Grocery Manufacturers of America, http://www.gmabrands.com/ publicpolicy/docs/Correspondence.cfm?DocID=862&, accessed 17/05/03.

Grocery Manufacturers of America News Release (2003) *More 'Stringent and Compre-hensive' System Required for Food Safety Supply*, Washington, D.C.: Grocery Manufacturers of America, http://www.gmabrands.com/news/docs/News Release.cfm?DocID=1134, accessed 17/05/03.

Grove-White, R. (2001) 'New Wine, Old Bottles? Personal Reflections on the New Biotechnology Commissions', *The Political Quarterly*, volume 72, no. 4: 466–472.

Gurian-Sherman, D. (2002) *Holes in the Biotech Safety Net*, Washington, D.C.: Center for Science in the Public Interest.

Gutting, G. (1989) *Michel Foucault's Archaeology of Scientific Reason*, Cambridge: Cambridge University Press.

Haas, P. (1992) 'Introduction: Epistemic Communities and International Policy Co-ordination' *International Organisation*, volume 46, no. 1: 1–35.

Habermas, J. (1971) *Towards a Rational Society*, London: Heinemann.

Habermas, J. (1982) 'A Reply to My Critics' in Thompson, J. and Held, D. (eds) *Habermas: Critical Debates*, London: Macmillan, pp. 219–283.

Habermas, J. (1996) *Between Facts and Norms*, Cambridge: Polity Press.

Haerlin, B. (1990) 'Genetic Engineering in Europe' in Wheale, P. and McNally, R. (eds) *The Bio Revolution: Cornucopia or Pandora's Box?*, London: Pluto Press, pp. 253–261.

Hajer, M. (1995) *The Politics of Environmental Discourse*, Oxford: Clarendon Press.

Hall, P. and Taylor, R. (1996) 'Political Science and the Three New Institution-alisms', *Political Studies*, volume XLIV, pp. 936–957.

Hall, P. and Taylor., R. (1998) 'The Potential of Historical Institutionalism: A Response to Hay and Wincott', *Political Studies*, volume XLVI, pp. 958–962.

Hannigan, J. (1995) *Environmental Sociology – A Social Constructionist Perspective*, London: Routledge.

Harhoff, D., Regibeau, P. and Rockett, K. (2001) 'Genetically Modified Food – Eval-uating the Economic Risks', *Economic Policy*, October, issue no. 33, pp. 263–299.

Hari, J. (2002) 'Whatever Happened to No Logo?', *New Statesman*, 11/11/02, pp. 20–22.

Harris, P. (1991), 'Biotechnology and the Organic Movement', *New Farmer and Grower*, issue no. 30, p. 29.

Hay, C. (2002) *Political Analysis – A Critical Introduction*, Basingstoke, Hampshire: Palgrave.

Hay, C. and Rosamond, B. (2002) 'Globalisation, European Integration and The Discursive Construction of Economic Imperatives', *Journal of European Public Policy*, volume 9, no. 2, pp. 147–167.

Hay, C. and Wincott, D. (1998) 'Structure, Agency and Historical Institutionalism', *Political Studies*, volume 46, no. 5, pp. 951–957.

Hayward, T. (1995) *Ecological Thought: An Introduction*, Oxford: Polity Press.

Heller, C. (2002) 'From Scientific Risk To Paysan Savoir-Faire: Peasant Expertise in the French and Global Debate Over GM Crops', *Science as Culture*, volume 11, no. 1.

Henderson, M. (2002) 'New GM Rice Could Transform the Fight Against Famine', *The Times*, November 26th, p. 14.

Henke, D. (1997) 'Fears Over "Killer" Crops', *Guardian*, December 18th, p. 1.

Hinscliff, G. (1999) 'Sainsbury and the GM Web', *Daily Mail*, February 22nd, p. 17.

House of Commons Hansard Debates (1999) February 3rd pt 22, Column 126.

Howarth, D. (1995) 'Discourse Theory' in Marsh, D. and Stoker, G. (eds) *Theory and Methods in Political Science*, Basingtoke, Hampshire: Macmillan.

Howarth, D. (2000) *Discourse*, Buckingham: Open University Press.

Huang, J., Rozelle, S., Pray, C. and Qinfang, W. (2002) 'Plant Biotechnology in China', *Science*, volume 295, no. 5555 (January 25th) pp. 674–677.

Huxham, M. (2000) 'Science and the Search for Truth' in Huxham, M. and Sumner, D. (eds) *Science and Environmental Decision Making*, Harlow, Essex: Prentice Hall, pp. 1–32.

Huxham, M. and Sumner, D. (2000) *Science and Environmental Decision Making*, Harlow, Essex: Prentice Hall.

Huxley, A. (1950) *Brave New World*, London: Chatto and Windus.

Irwin, A. (1995) *Citizen Science*, London: Routledge.

Jasanoff, S. (1990) *The Fifth Branch: Science Advisors as Policymakers*, Cambridge, Massachusetts: Harvard University Press.

Jasanoff, S. (1995) 'Product, Process or Programme; Three Cultures and the Regulation of Biotechnology' in M. Bauer (ed.) *Resistance to New Technology*, Cambridge; Cambridge University Press, pp. 311–334.

Jenkins-Smith, H. and Sabatier, P. (1993) 'The Study of Public Policy Processes' in Sabatier, P. and Jenkins-Smith, H. (eds) *Policy Change and Learning – An Advocacy Coalition Approach*, San Francisco: Westview Press.

Johnson, B. (2003) 'Degraded Land', letter to *New Scientist* no. 2376, January 4th, p. 23.

Joint Meeting of the Food Advisory Committee and The Veterinary Medicine Advisory Committee (1994) volume II, November 3rd, Arlington, Virginia: Department of Health and Human Services, Public Health Service, Food and Drug Administration.

Jordan, A (2002) 'Introduction: European Union Environmental Policy – Actors, Institutions and Policy Processes' in A. Jordan (ed.) *Environmental Policy in the European Union*, London: Earthscan, pp. 1–12.

Judge, D. (2002) 'Predestined to Save the Earth: The Environment Committee of the European Parliament' in A. Jordan (ed.) *Environmental Policy in the European Union*, London: Earthscan, pp. 120–140.

Kingdon, J. (1995) *Agendas, Alternatives and Public Policies*, 2nd edition, New York: HarperCollins.

Kishaw, G. and Shewmaker, C. (1999) 'Biotechnology: Enhancing Human Nutrition in Developing and Developed Worlds', 'Plants and Population: Is there Time?, Proceedings National Academy of Sciences, volume 96, pp. 5968–5972, May 1999, Colloquium Paper, Washington, D.C.: National Academies Press, http://books.nap.edu/books/0309064279/html/5968.html, accessed 29/05/03.

Klein, J. (2002) *The Natural*, New York: Broadway Books.

Krenzler, H. and MacGregor, A. (2000) 'GM Food: The Next Major Transatlantic Trade War?', *European Foreign Affairs Review*, no. 5, pp. 287–316.

Krimsky, S. and Murphy, N. (2002) 'Biotechnology at the Dinner Table: FDA's Oversight of Transgenic Food', *ANNALS*, American Academy of Political and Social Science, p. 584.

Krimsky, S. and Wrubel, R. (1996) *Agricultural Biotechnology and the Environment*, Chicago: University of Illinois Press.

Kuhn, T. (1970) *The Structure of Scientific Revolutions*, Chicago: University of Chicago Press.

Lake, G. (1991) 'Scientific Uncertainty and Political Regulation: European Legislation on the Contained Use and Deliberate Release of Genetically Modified (Micro) Organisms', *Project Appraisal*, volume 6, no. 1, pp. 7–15.

Leach, G. (1969) 'How the Biologists Started a Scare', *Observer*, November 30th, p. 9.

Lean, G. (2002) 'GM Crops are Breeding with Plants in the Wild', *Independent*, December 29th, p. 1.

Lee, K. (1993) *Compass and Gyroscope*, Washington, D.C.: Island Press.

Lee, T. (2000) *Reconciling Lay and Expert Evaluations of the Riskiness of Hazardous Technologies*, paper prepared for the OECD Workshop on 'The Adoption of Technologies for Sustainable Farming Systems', Wageningen, July 4th–7th.

Legislative Observatory (1992) 'Regulation (EC) No. 258/97 of the European Parliament and of the Council of 27 January 1997 Concerning Novel Foods and Novel Food Ingredients', and accompanying notes, http://www. db.europarl.eu.int/oeil/oeil.Res112, accessed 24/06/03.

Levidow, L. (2001) 'Precautionary Uncertainty: Regulating GM Crops in Europe', *Social Studies of Science*, volume 31, no. 6, pp. 842–874.

Levidow, L. and Carr, S. (2000) 'Normalizing Novelty: Regulating Biotechnological Risk at the U.S. EPA', *11 Risk: Health, Safety & Environment 9* (Winter 2000), pp. 9–34.

Levidow, L., Carr, S., von Schomberg, R. and Wield, D. (1996) 'Regulating Agricultural Biotechnology in Europe: Harmonization Difficulties, Opportunities, Dilemmas', *Science and Public Policy*, volume 23, no. 3, pp. 135–157.

Liberatore, A. (1995) 'The Social Construction of Environmental Problems' in Blowers, A. and Glasbergen, P. (eds) *Perspectives on Environmental Problems*, London: Arnold, pp. 59–84.

Litfin, K. (1994) *Ozone Discourses*, New York: Columbia University Press.

Lofstedt, R. and Vogel, D. (2001) 'The Changing Character of Regulation: A Comparison of Europe and the United States', *Risk Analysis*, volume 21, no. 3, pp. 399–405.

Losey, J., Raynor, L. and Carter, M. (1999) 'Transgenic Pollen Harms Monarch Larvae', *Nature*, no. 399, p. 214.

Lowndes, V. (2002) 'The Institutional Approach' in Marsh, D. and Stoker, G. (eds) *Political Theory and Methods*, London: Palgrave Macmillan, pp. 90–108.

Lynch, D. and Da Ros, J. (2001) *Science and Public Participation in Regulating Genetically-Engineered Food: French and American Experiences*, Paper to workshop organised by European Institute of Business Administration (INSEAD) and Berkeley University of California at INSEAD, Fontainebleau, June 8th and 9th.

MacMillan, T. (2002) 'Governing Interests – Science, Power and Scale in the rBST Controversy', PhD thesis, Manchester: School of Geography, University of Manchester.

McHughen, A. (2000) 'The Regulation of GM Foods – Who Represents the Public Interest?', *International Journal*, Autumn 2000, LV, pp. 624–632.

Majone, G. (1992) 'Regulatory Federalism in the European Community', *Environment and Planning C: Government and Policy*, volume 10, no. 3, pp. 299–316.

Manning, F. (2000) 'Biotechnology: A Scientific Perspective' in Russell, A. and Vogler, J. (eds) *The International Politics of Biotechnology*, Manchester: Manchester University Press, pp. 13–29.

March, J. and Olsen, J. (1989) *Rediscovering Institutions – The Organizational Basis of Politics*, New York: The Free Press.

Martineau, B. (2001) *First Fruit*, New York: McGraw-Hill.

Marsh D. and Rhodes, R. (1992) *Policy Networks in British Government*, Oxford: Oxford University Press.

Mayer, S. (2000) 'Genetic Engineering in Agriculture' in Huxham, M. and Sumner, D. (eds) *Science and Environmental Decision Making*, Harlow, Essex: Prentice Hall, pp. 94–117.

McGlade, J.M. (1993) 'Ideal Speech Situation' online definition attributed to McGlade, J.M. (1993) 'Governance of fisheries and aquaculture', final report, contract FAR MA 3. 757, DGXIV European Union http://homepages. which.net/~gk.sherman/baaaaacs.htm, accessed 10/12/02.

Meikle, J. (1999) 'GM Measures Scorned', *Guardian*, May 22nd, p. 1.

Mellon, M. (1988) *Biotechnology and the Environment*, Washington, D.C.: National Wildlife Federation.

Mellon, M. (1998) 'UCS Introduction' in Mellon, M. and Rissler, J. (eds) *Now Or Never – Serious New Plans to Save a Natural Pest Control*, Washington, D.C.: Union of Concerned Scientists, pp 1–12.

Meyet, S. (2002) 'Political Theory After/Beyond Foucault: An Interview with Mark Bevir', *Post-Structuralism and Radical Politics Newsletter*, issue 4.

Midden, J. (1995) 'Direct Participation in Macro-Issues: A Multiple Group Approach' in Renn, O., Webler, T. and Wiedemann, P. (eds) *Fairness and Competence in Citizen Participation*, Boston: Kluwer, pp. 305–320, http://homepages. gold.ac.uk/psrpsg/news/r4bevir.html, accessed 26/11/02.

Mikl, M. and Torgersen, H. (1996) 'Austria's Biotechnology Regulation: from "Virtual Releases" to Public Protest', *Science and Public Policy*, volume 23, no. 3, pp. 195–200.

Monsanto (2002) 'Scientists Publish Evaluation of BiotechCrops' http://www. monsanto.com/monsanto/layout/media/00/12-15-00.asp, accessed 26/10/02.

Mouffe, C. (1995) 'Post Marxism: Democracy and Identity' *Environment and Planning D: Society and Space*, volume 13, pp. 259–265.

Mumpower, J. (1995) 'Dutch Study Groups Revisited', in Renn, O., Webler, T. and Wiedemann, P. (eds) *Fairness and Competence in Citizen Participation*, Boston: Kluwer, pp. 321–338.

National Academy of Sciences (NAS) (2002a) *Genetically Modified Pest-Protected Plants – Science and Regulation*, Washington, D.C.: National Academies Press.

National Academy of Sciences (NAS) (2002b) *Environmental Effects of Transgenic Plants: The Scope and Adequacy of Regulation*, Washington, D.C.: National Academies Press, http://bob.nap.edu/books/0309082633/html/, accessed 29/05/03.

Negin, E. (1996) 'The ALAR 'Scare' Was For Real', *Columbian Journalism Review*, October/November 1996, http://www.cjr.org/year96/5/alar.asp, accessed 13/04/03.

Nestle, M. (2002) *Food Politics: How the Food Industry Influences Nutrition and Health*, Berkeley, California: University of California Press.

Nestle, M. (2003) *Safe Food – Bacteria, Biotechnology and Bioterrorism*, Berkeley, California: University of California Press.

Netlink (1996) 'Austria Bans Frankencorn', http://www.netlink.de/gen/Zeitung/1223.htm.

New Scientist Editorial (2000) 'We Can't Ignore the Taco Fiasco. Next Time it Could Be Serious', *New Scientist*, October 7th, p. 3.

Office of Science and Technology Policy (OSTP) (1986) *Coordinated Framework for Regulation of Biotechnology*, 51 FR 23302, http://web.lexis-nexis.com/cong-comp/docu … AA&_md5=b3ce0d0eac750cbed16ad6e5b3615896, accessed at Johns Hopkins University Library 01/03/03.

Official Journal of the European Communities (1990a) Council Directive of April 23rd on the Deliberate Release into the Environment of Genetically Modified Organisms, (90/220/EEC), *Official Journal* L117, 08/05/90, pp. 15–27.

Official Journal of the European Communities (1990b) Minutes of the Sitting of the European Parliament, *Official Journal* C96, 14/03/90, pp. 88 and 90, cited by Lake (1991), p. 12.

Official Journal of the European Communities (1997) Regulation (EC) No 258/97 of the European Parliament and of the Council of January 27th Concerning Novel Foods and Novel Ingredients, *Official Journal* L043, 14/02/97, pp. 1–6.

Official Journal of the European Communities (1999) Proposal for a European Parliament and Council Directive Amending 90.220/EEC on the Deliberate Release into the Environment of Genetically Modified Organisms, *Official Journal* C150, 28/05/99, pp. 363–380.

Official Journal of the European Communities (2001) Directive 2001/18/EC of the European Parliament and the Council of March 12th on the Deliberate Release into the Environment of Genetically Modified Organisms and Repealing Council Directive 90/220/EEC, *Official Journal* L106, 17/04/01, pp. 1–38.

Official Journal of the European Union (2003) Regulation (EC) No. 1829/2003 of the European Parliament and of the Council of September 22nd on Genetically Modified Food and Feed, *Official Journal of the European Communities* L268, 18/10/03 pp. 1–23.

OLR/Opinion Learning Research (2003) 'Should GM Food be Available to Buy in the UK?' London: Food Standards Agency, http://www.foodstandards.gov.uk/multimedia/pdfs/gmcitjuryfinalrep.pdf, accessed 25/07/03.

O'Riordan, T. (2000) 'Environmental Science on the Move' in O'Riordan, T.

(ed.) *Environmental Science for Environmental Management*, Harlow, England: Prentice Hall, pp. 1–28.

O'Riordan, T. and Cameron, J. (eds) (1994) *Interpreting the Precautionary Principle*, London: Earthscan.

O'Riordan, T. and Cobb, D. (2001) 'Assessing the Consequences of Converting to Organic Agriculture', *Journal of Agricultural Economics*, volume 52, no. 1, pp. 22–35.

Orson, J. (2002) *Gene Stacking in Herbicide Tolerant Oilseed Rape: Lessons for the North Amercian Experience*, Peterborough: English Nature Report no. 443.

Ostrom, E. (1990) *Governing the Commons, the Evolution of Institutions for Collective Action*, Cambridge: Cambridge University Press.

Peters, G. (1998) *Comparative Politics – Theory and Methods*, Basingstoke, Hampshire: Macmillan.

Petts, J., Horlick-Jones, T. and Murdock, G. (2001) *Social Amplification of Risk: the Media and the Public*, London: Health and Safety Executive.

Pew Initiative on Food and Biotechnology (2002) *Pharming The Field – A Look at the Benefits and Risks of Bioengineering Plants to Produce Pharmaceuticals*, Washington, D.C.: Pew Initiative/FDA/USDA.

Phillips, M.W. (Lord) (2000a) *The BSE Inquiry: The Report* – volume 1: Findings and Conclusions, Executive Summary of the Report of the Inquiry, 4. Assessment of Risks Posed by BSE to Humans, London: Ministry of Agriculture, Fisheries and Food, p. 2, http://www.bseinquiry.gov.uk/report/volume1/toc.htm, accessed 30/12/02.

Phillips, M.W. (Lord) (2000b) *The BSE Inquiry: The Report* – volume 1: Findings and Conclusions, Executive Summary of the Report of the Inquiry, 1. Key Conclusions, London: Ministry of Agriculture, Fisheries and Food, p. 2, http://www.bseinquiry.gov.uk/report/volume1/toc.htm, accessed 30/12/02.

Phillips, M.W. (Lord) (2000c) *The BSE Inquiry: The Report* – volume 1: Findings and Conclusions, 14. Lessons To Be Learned, London: Ministry of Agriculture, Fisheries and Food, para 1294, http://www.bseinquiry.gov.uk/report/volume1/toc.htm, accessed 30/12/02.

Pickering, A. (1992) 'From Science as Knowledge to Science as Practice' in Pickering, A. (ed.) *Science As Practice and Culture*, Chicago: University of Chicago Press.

Pierson, P. (1993) 'When Effect Becomes Cause: Policy Feedback and Political Change', *World Politics*, volume 45, no. 4, pp. 595–628.

Porritt, J. (1984) *Seeing Green*, Oxford: Blackwell.

Pusztai, A. (2002) 'GM Food Safety: Scientific and Institutional Issues', *Science As Culture*, volume 11, no. 1, pp. 69–92.

Quist, D. and Chapela, I. (2001) 'Transgenic DNA Introgressed into Traditional Maize Landraces in Mexico', *Nature*, issue 414, pp. 541–543.

Randall, E. (2001) 'Uneasy Relationships: Learning the Lesson and Sharing the Load', Ed Randall and Policy Library, London: Goldsmith's College, http://www.policylibrary.com/Essays/RandallEFARisk/EFArisk4.htm, accessed 03/07/03.

Reilly, J. and Miller, D. (1997) 'Scaremonger and Scapegoat? The Role of the Media in the Emergence of Food as a Social Issue' in Caplan, P. (ed.) 'Approaches to Food, Health and Identity' in *Food, Health and Identity*, London: Routledge, pp. 234–251.

Renn, O., Webler, T. and Wiedemann, P. (1995) *Fairness and Competence in Citizen Participation*, Boston: Kluwer.

Reuters (1996) 'Austrian Opposition to Genetic Foods Gathers Pace', Vienna: Reuters, http://www.netlink.de/gen/Zeitung/1203a.htm, accessed 26/06/03.

Reuters (1997) 'Top French Scientist Resigns over Gene Maize Row', Paris: Reuters, http://www.netlink.de/gen/Zeitung/970213.htm, accessed 26/06/03.

Rich, A. (1997a) 'La lecon de la vache folle n'a pas suffi', *Le Soir,* January 27th, p. 1.

Rich, A. (1997b) 'Actualite Internationale', *Le Soir,* April 9th.

Rifkin, J. (1983) *Algeny,* New York: Viking Press.

Rifkin, J. (1998) *The Biotech Century,* London: Victor Gollancz.

Riley, P. and Bebb, A. (1997) ' "Food and Biotechnology", Change the World – Bulletin for Local Groups', London: Friends of the Earth.

Rissler, J. and Mellon, M. (1996) *The Ecological Risks of Engineered Crops,* Cambridge, Massachusetts: MIT Press.

Roberts, M. (2002) *Spanish Farmers Seen Reaping Rewards from GM Maize,* London: Reuters English News Service.

Rogers, M. (2001) 'Commentary on Lofstedt and Vogel' in *Risk Analysis,* volume 21, no. 3, pp 412–415.

Rorty, R. (1986) 'Foucault and Epistemology' in Couzens Hoy, D. (ed.) *Foucault: A Critical Reader,* Oxford: Blackwell, pp. 41–49.

Roy, A. and Joly, P.-B. (2000) 'France: Broadening Precautionary Expertise?', *Journal of Risk Research,* volume 3, no. 3, pp. 247–254.

Royal Commission on Environmental Pollution (RCEP) (1989) *Thirteenth Report – The Release of Genetically Engineered Organisms to the Environment,* London: HMSO.

The Royal Society (2002) *Genetically Modified Plants for Food Use and Human Health – an Update,* London: The Royal Society, www.royalsoc.ac.uk, accessed 10/02/03.

Russell, A. and Vogler, J. (2000) *The International Politics of Biotechnology,* Manchester: Manchester University Press.

Sabatier, P. and Jenkins-Smith, H. (1993) *Policy Change and Learning – An Advocacy Coalition Approach,* San Francisco: Westview Press.

SAGB (1990) Senior Advisory Group on Biotechnology (later renamed 'EuropaBio'), *Community Policy for Biotechnology: Economic Benefits and European Competitiveness,* Brussels: European Chemical Industry Council, cited by Levidow (2001a: 849).

Sanders, D. (1995) 'Behaviouralism' in Marsh, D. and Stoker, G. (eds) *Political Theory and Methods,* London: Palgrave Macmillan, pp. 45–64.

Sassatelli, R. and Scott, A. (2001) 'Novel Food, New Markets and Trust Regimes', *European Societies,* volume 3, no. 2, pp. 213–244.

Save Biodiversity! (2002) unsigned, *Save The Monarch Butterfly,* http://www.environmentalaction.net/monarch/monarch_epa.html, accessed 22/10/02.

Scharpf, F. (1988) 'The Joint-Decision Trap: Lessons from German Federalism and European Institutions', *Public Administration,* volume 66, no. 3, pp. 239–278.

SCF (1997) EU Scientific Committee on Food, *Opinion on the Additional Information from the Austrian Authorities Concerning the Marketing of Ciba Geigy Maize* (expressed on March 21st), Brussels: DG Health and Consumer Protection, http://europa.eu.int/comm/food/fs/sc/oldcomm7/out01_en.html, accessed 01/07/03.

Scott, R. (1995) *Institutions and Organisations,* Thousand Oak, California: Sage.

SCP (1998) EU Scientific Committee on Plants, *Opinion of the Scientific Committee on Plants Regarding Submission for Placing on the Market of Genetically Modified High Amylopectin Potato Cultivars apriori and apropos Notified by Avebe* (Opinion Adopted on October 2nd), Brussels: DG Health and Consumer Protection, http://europa.eu.int/comm/food/fs/sc/scp/out24_en.html, accessed 23/09/02.

Shackley, S. (1993) *Regulating the New Biotechnology in Europe*, D. Phil thesis, Falmer, Brighton: Science Policy Research Unit, University of Sussex.

Shackley, S., Levidow, L. and Tait, J. (1992) 'Contending Rationalities and Regulatory Politics: the Case of Biotechnology', unpublished manuscript archived in Manchester School of Management, University of Manchester Institute of Science and Technology (UMIST).

Shelley, M. (1994) *Frankenstein, Or The Modern Prometheus*, Oxford: Oxford University Press.

Shiva, V. (2000) *Tomorrow's Biodiversity*, London: Thames & Hudson.

Slovic, P., Fischoff, B. and Lichtenstein, S. (1980) 'Facts and Fears: Understanding Perceived Risk' in Schwing, R. and Albers, W. (eds) *Societal Risk Assessment: How Safe is Safe Enough*, New York: Plenum Press, cited by Lee (2000).

Slow Food FAQ (2003) Slow Food Movement, Slowfood.com, accessed 24/06/03.

Smith, M. (1991) 'From Policy Community to Issue Network: Salmonella in Eggs and the New Politics of Food', *Public Administration*, volume 69, no. 2, pp. 235–255.

Snow, C.P. (1959) *The Two Cultures and the Scientific Revolution*, Cambridge; Cambridge University Press.

Soil Association (2001) Fact Sheet: Organic Farming in Europe, Bristol: Soil Association.

Southwood, R. (1989) *Report of the Working Party on Bovine Spongiform Encephalopathy*, London: HMSO, cited by Reilly, J. and Miller, D. 'Scaremonger and Scapegoat? The Role of the Media in the Emergence of Food as a Social Issue' in Caplan, P. (ed.) (1997) 'Approaches to Food, Health and Identity' in *Food, Health and Identity* London: Routledge, pp. 234–251.

Spetnak, C. and Capra, F. (1985) *Green Politics*, London: Paladin Grafton.

Steinmo, S., Thelen, K. and Longstreth, F. (1993) 'Structuring Politics', Cambridge: Cambridge Univeristy Press.

Stratton, D. (1977) 'The Genetic Engineering Debate', *Ecologist*, volume 7, no. 10, pp. 381–388.

Terlouw, J.C., de Boois, H.M., Dorrestein, R., Galjaard, H., Kok, F.J., van der Laan-Veraart, M.D.A.M., de la Rive, L., Scheffer, H.C. and Seydel, E.R. (2002) *Eten en Genen, een publiek debat over biotechnologie en voedsel, verslag van de tijdelijke commissie biotechnologie en voedsel*, Den Haag: ministerie van LNV – unofficial translation by Koen Jonkers.

Toke, D. (1999) 'Epistemic Communities and Environmental Pressure Groups', *Politics*, volume 19, no. 2, pp. 97–102.

Toke, D. (2000a) *Green Politics and Neo-Liberalism*, London: Macmillan.

Toke, D. (2000b) 'Policy Network Creation: the Case of Energy Efficiency', *Public Administration*, volume 78, no. 4.

Toke, D. (2002a) 'Ecological Modernisation and GM Food', *Environmental Politics*, volume 11, no. 3, pp. 145–163.

Toke, D. (2002b) 'Wind Power in UK and Denmark: Can Rational Choice Help Explain Different Outcomes?' *Environmental Politics*, volume 11, no. 4, pp. 83–100.

Toke, D. (2003) 'Research Note: A Comparative Study of the Politics of GM Food and Crops', *Political Studies*, volume 51, no. 3.

Toke, D. and Marsh, D. (2003) 'Policy Networks and the GM Crops Issue: Assessing the Utility of the Dialectical Model of Policy Networks', *Public Administration*, forthcoming.

USDA Agricultural Research Service (2003) 'Notice of Establishment of the Advisory Committee on Biotechnology and 21st Century Agriculture and Notice of Appointment of Committee Members', Federal Register Online: April 15th (volume 68, no. 72), http://a257.g.akamaitech.net/7/257/2422/14mar20010800/edocket.access.gpo.gov/2003/03-9173.htm, accessed 06/06/03.

USDA News Release (2003a) *Nobel Peace Laureate Urges Adoption of Agricultural Technology to Feed the Developing World*, Washington, D.C.: United States Department of Agriculture, release number 0203.03, June 25th.

USDA News Release (2003b) *U.S. and Cooperating Countries File WTO Case Against EU Moratorium and Biotech Foods and Crops*, Washington, D.C.: United States Department of Agriculture, release number 0156.03, May 13th.

Van der haegen, T. (2002) *The Looming US–EU Conflict Over Plant Biotechnology and Trade* – Speech to Cato Policy Forum, Washington, D.C. September 25th, Washington, D.C.: The European Union in the US, http://www.eurunion.org/News/speeches/2002/020925tvdh.htm#Annex2, accessed 15/05/03.

Vidal, J. (1998a) 'Genetically Altered Crops "Could Wipe Out Farmland Birds"', *Guardian*, July 8th, p. 11.

Vidal, J. (1998b) 'Public "Wants Labels on Genetically Modified Food"', *Guardian*, June 4th, p. 12.

Vogel, D (2001) *Ships Passing in the Night: GMOs and the Politics of Risk Regulation in Europe and the United States*, paper to Workshop organised by European Institute of Business Administration (INSEAD) and Berkeley University of California at INSEAD, Fontainebleau, June 8th and 9th.

Weale, A. (1996) 'Environmental Rules and Rule-Making in the European Union', *Journal of European Public Policy*, volume 3, no. 4, pp. 594–611.

Weale, A. (2002) 'European Environmental Policy by Stealth: the Dysfunctionality of Functionalism, in A. Jordan (ed.) *Environmental Policy in the European Union*, London: Earthscan, pp. 329–347.

Webler, T. (1995) '"Right" Discourse in Citizen Participation: An Evaluative Yardstick' in Renn, O., Webler, T. and Wiedemann, P. (eds) *Fairness and Competence in Citizen Participation*, Boston: Kluwer, pp. 35–86.

Wells, H.G. (1928) *Men Like Gods*, London: Collins.

Wells, S. (2003) *GM Crop Debacle*, *Daily Mail*, July 22nd, p. 10.

Wynne, B. (1992a) 'Uncertainty and Environmental Learning: Reconceiving Science and Policy in the Preventative Paradigm', *Global Environmental Change*, volume 2, no. 2 (June) 111–127, cited by Levidow 2001a.

Wynne, B. (1992b) 'Carving Out Science (And Politics) in the Regulatory Jungle' (Essay Review), *Social Studies of Science*, volume 1, no. 3, pp. 745–758, cited by Hajer 1995: 139.

Wynne, B. (1996) 'May the Sheep Safely Graze? A Reflexive View of the Expert-Lay Knowledge Divide, in Lash, S., Szerszynski, B. and Wynne, B. (eds) *Risk Environment and Modernity*, London: Sage.

Index